Studies on Amanita (Amanitaceae) from Andean Colombia

Rodham E. Tulloss

Clark L. Ovrebo

Roy E. Halling

WITHDRAWN

AF697224

EMOIRS OF THE NEW YORK BOTANICAL GARDEN VOLUME 66

Memoirs of The New York Botanical Garden

ADVISORY BOARD

PATRICIA K. HOLMGREN, Director
of the Herbarium
The New York Botanical Garden

JAMES L. LUTEYN, Senior Curator
The New York Botanical Garden

SCOTT A. MORI, Senior Curator
The New York Botanical Garden

EDITORIAL BOARD

WILLIAM R. BUCK, Editor
The New York Botanical Garden
Bronx, New York 10458

WM. WAYT THOMAS, Associate Editor
The New York Botanical Garden
Bronx, New York 10458

H. DAVID HAMMOND, Associate Editor
The New York Botanical Garden
Bronx, New York 10458

MELINDA F. DENTON (1986–1991), Herbarium, Department of Botany, University of Washington, Seattle, Washington 98195 U.S.A.

THOMAS F. DANIEL (1991–1996), Department of Botany, California Academy of Sciences, Golden Gate Park, San Francisco, California 94118 U.S.A.

The MEMOIRS are published at irregular intervals in issues and volumes of various sizes and are designed to include results of original botanical research by members of the Garden's staff, or by botanists who have collaborated in one or more of the Garden's research programs. Ordinarily only manuscripts of fifty or more typewritten pages will be considered for publication.

Manuscripts should be submitted to the Editor. For further information regarding editorial policy and instructions for the preparation of manuscripts, address the Editor.

Orders for past and forthcoming issues and volumes should be placed with:
Scientific Publications Department
The New York Botanical Garden
Bronx, New York 10458-5126 U.S.A.

THIS PUBLICATION IS PRINTED ON ACID-FREE PAPER.

A10207 053908

Studies on Amanita (Amanitaceae) from Andean Colombia

Rodham E. Tulloss

Clark L. Ovrebo

Roy E. Halling

The New York Botanical Garden
Bronx, New York 10458, U.S.A.
Issued: 13 January 1992

MEMOIRS OF THE NEW YORK BOTANICAL GARDEN VOLUME 66

Front Cover Illustration from "Descrizione dei funghi mangerecci piu comuni dell'Italia e de' velenosi che possono co' medesimi confonderesi," by Carlo Vittadini. 1835, Milano. 364 p.

QK
1
.N525
v.66

Copyright © 1992

The New York Botanical Garden

Published by
The New York Botanical Garden
Bronx, New York 10458-5126

International Standard Serial Number 0071-5794

Library of Congress Cataloging-in-Publication Data

Tulloss, Rodham E.
Studies on Amanita (Amanitaceae) from Andean Colombia / Rodham E. Tulloss, Clark L. Ovrebo, Roy E. Halling.
p. cm. — (Memoirs of the New York Botanical Garden ; v. 65)
Includes bibliographical references and index.
ISBN 0-89327-371-6
1. Amanita—Colombia. 2. Amanita—Andes. I. Ovrebo, Clark L. II. Halling, Roy E. III. Title. IV. Series.
QK1.N525 vol. 65a
[QK629.A53]
581 s—dc20 91-37518
[589.2'22] CIP

Copyright © 1992 The New York Botanical Garden
International Standard Book Number 0-89327-371-6

JANUARY 1992 MEMOIRS OF THE NEW YORK BOTANICAL GARDEN **66:** 1–46

Studies on Amanita (Amanitaceae) from Andean Colombia

Rodham E. Tulloss, Clark L. Ovrebo, Roy E. Halling

Contents

Abstract

Rodham E. Tulloss (P. O. Box 57, Roosevelt, NJ 08555-0057, U.S.A.) **Clark L. Ovrebo** (Department of Biology, University of Central Oklahoma, Edmond, OK 73034, U.S.A.) and **Roy E. Halling** (Cryptogamic Herbarium, The New York Botanical Garden, Bronx, NY 10458-5126, U.S.A.). Studies on *Amanita* (Amanitaceae) from Andean Colombia. Mem. New York Bot. Gard. **66:** 1–46. 1992.—We review collections of *Amanita* of Colombia largely from *Quercus humboldtii* forests. The number of taxa of *Amanita* recorded for Colombia is increased from six to thirteen. *Amanita advena, A. arocheae, A. aureomonile, A. brunneolocularis, A. colombiana, A. flavoconia* var. *inquinata, A. flavoconia* var. *sinapicolor, A. fuligineodisca, A. picea, A. sororcula*, and *A. xylinivolva* are described as new. We believe two of these (*A. arocheae* and *A. aureomonile*) are the first species of *Amanita* reported from a *Colombobalanus* forest. A type study of *A. humboldtii* is reported. With the exception of *A. flavoconia, A. humboldtii*, and *A. muscaria* var. *flavivolvata*, the species previously reported from Colombia are cases of misapplication of European names to a member of the preceding list. A key to the species of *Amanita* known from Colombia is provided.

Resumen

Revisamos colecciones de especies de *Amanita* para Colombia, en su mayor parte en bosques de *Quercus humboldtii*. El número de taxa conocidas de *Amanita* para Colombia ha sido aumentado de seis a trece. *Amanita advena, A. arocheae, A. aureomonile, A. brunneolocularis, A. colombiana, A. flavoconia* var. *inquinata, A. flavoconia* var. *sinapicolor, A. fuligineodisca, A. picea, A. sororcula* y *A. xylinivolva* se han descrito como nuevas. Creemos que dos de estas especies (*A. arocheae* y *A. aureomonile*) son las primeras reportadas para el género en bosques de *Colombobalanus*. Un estudio del tipo de *A. humboldtii* es reportado. Con la excepción de *A. flavoconia, A. humboldtii,* y *A. muscaria* var. *flavivolvata,* las especies reportadas previamente para Colombia son casos de conceptos erróneos aplicados a la anterior lista. Se provee una clave para las especies de *Amanita* conocidas en Colombia.

Key Words: Colombia, Costa Rica, Mexico, *Amanita, Quercus, Colombobalanus, Trigonobalanus*

Introduction

This study was originally motivated by collections of *Amanita* Pers. made in *Quercus humboldtii* Bonpl. forests during expeditions to Colombia by Halling and Ovrebo in 1986, by Halling in 1987, by Halling and Ana E. Franco-Molano in 1988 and by Franco-M. in 1990. These trips resulted in the discovery of a number of interesting taxa (*e.g.*, Halling & Ovrebo, 1987a, 1987b; Halling, 1989a, 1989b; Singer *et al.*, 1990). The Halling, Ovrebo, and Franco-M. collections of *Amanita* were supplemented in our study by the examination of material from F collected by G. M. Mueller and R. Singer and by a number of collections previously mentioned in the literature including the holotype of *A. humboldtii* Sing.—the only *Amanita* species previously described originally from *Q. humboldtii* forests.

The most recent review of taxa of *Amanita* known from Colombia is that of Pulido-O. (1983). She cited five taxa recorded in the previous literature: *Amanita gemmata sensu* Guzmán & Varela, *A. humboldtii*, "*A. inaurata*" *sensu auct.*, *A. muscaria* (L.:Fr.) Pers., and *A. rubescens* Pers.:Fr. Henao-M. and Tobón-A. (1984) reported on their collections from Antioquia. Of the preceding list, they omitted *A. humboldtii*, and they added an unnamed species that they determined to have affinity to *A. flavoconia* Atk. Hence, six *Amanita* taxa had been documented in Colombia. From the material we have examined, it is clear that that number can be more than doubled. We will also demonstrate that not all the names listed above have been correctly applied.

Materials and Methods

All anatomical studies were done on dried material. Microscopic examination of tissues was made in 10% NH_4OH or 2%−3% KOH. All tissues were hyaline to very pale yellow unless otherwise noted. Stains used were Congo red and eosin. Spores were examined in Melzer's reagent and 10% NH_4OH or 2%−3% KOH. Macrochemical spot tests were carried out when possible. Those tests for tyrosinase (using paracresol) and laccase (using syringaldazine) were conducted following the procedure of Marr (1979). In descriptions of marginal striations of the pileus, the abbreviation nR is used where n is the ratio of striation length to the radius of the pileus (R). Dimensions for spores are given using notation of the form $(a-)b-c(-d)$. The range $b-c$ contains a minimum of 90% of the measured values. If less than 100% of the measured values fall within the range $b-c$, then a (if such exists $< b$) represents the smallest value measured and d (if such exists $>c$) represents the largest value measured. **L** is used to mean "average length" for spores on a specimen by specimen basis. When more than one specimen has been reviewed, the overall average is indicated by the addition of an apostrophe in this manner: **L'**. **W** is used to mean "average width" for spores on a specimen by specimen basis; Q is used to mean "length/width ratio" of a spore; **Q** is used to mean "average of length/width ratio" for spores on a specimen by specimen basis. **W'** and **Q'** are defined analogously to **L'**. We follow Bas (1974) in the following abbreviation: [*n*/*m*/*p*] shall mean "*n* spores measured from *m* specimens out of *p* collections."

We use the phrase "filamentous, undifferentiated hyphae" to describe the simple, mostly thin-walled hyphae of the tissue examined which lack refractive contents. Color codes of the form "11D2" are from Kornerup and Wanscher (1983). Color codes of the form "7.5YR 8/6" are from Munsell Color (1975). Color codes of the form "81. d. gy. y Br" are from Kelley (1965). Color names used by Singer in his notes on Colombian material came from Maerz and Paul (1930) and are followed in the text by the letters "MP" in parentheses.

With two exceptions, herbarium codes used follow Holmgren *et al.* (1990); the private herbarium of Tulloss is encoded by "RET" and the herbarium of the Dirección General de Flora y Fauna Silvestres, Facultad de Ciencias, U.N.A.M., México, by "SEDUE."

Taxonomic Treatment

Corner and Bas (1962) provided a key to two subgenera and six sections of *Amanita* which was refined by Bas (1969). A variant key was provided by Jenkins (1986). The key from Bas (1969) is duplicated here. The only section not represented in the known Colombian mycota is section *Amidella*.

1. Spores amyloid. Margin of pileus rarely radially sulcate. Lamellulae often attenuate .... Subgenus *Lepidella*

  2. Universal veil pulverulent or breaking up into flocks, warts, scales, patches, belts or crusts on pileus and stipe; sometimes disappearing completely from base of stipe. If bulb of stipe marginate, then universal veil floccose or forming conical warts, at least at center of pileus, and never provided with (sub)membranous outer layer.

    3. Margin of pileus not appendiculate. Pileipellis often deeply colored. Spores globose to ellipsoid, mostly < 10 μm, rarely up to 12 μm long. Partial veil membranous, rarely fugacious .... Section *Validae*

    3. Margin of pileus appendiculate. Pileipellis rarely deeply colored. Spores globose to bacilliform, rather often > 10 μm long. Partial veil membranous to friable .... Section *Lepidella*

  2. Universal veil circumscissile, limbate or saccate.

    4. Universal veil circumscissile or limbate, breaking up into submembranous, rarely pulverulent or floccose patches on pileus and leaving a slight, (sub)membranous limb or fragments of a limb at base of stipe, or else bulb distinctly marginate. Remnants of universal veil never conical warts.

      5. Margin of pileus appendiculate. Spores broadly ellipsoid to bacilliform. Partial veil membranous, friable or fugacious .... Section *Lepidella*

      5. Margin of pileus not appendiculate. Spores globose to subglobose. Partial veil membranous .... Section *Phalloideae*

4. Universal veil saccate, forming a membranous sac at base of stipe and only occasionally one or a few membranous patches on pileus; but sometimes inner layer of universal veil friable and forming scales, patches or powder on pileus.

6. Pileipellis often distinctly colored. Spores globose to ellipsoid. Partial veil membranous. Universal veil without friable inner layer. Lamellae not strongly darkening after drying. .... Section *Phalloideae*

6. Pileipellis usually white to brownish. Spores broadly ellipsoid to bacilliform. Partial veil often friable. Universal veil sometimes with friable inner layer. Lamellae often darkening strongly on drying. None known from South America .... Section *Amidella*

1. Spores inamyloid. Margin of pileus mostly radially sulcate. Lamellulae nearly always truncate .... Subgenus *Amanita*

7. Stipe with bulbous base. Universal veil usually friable, sometimes limbate .... Section *Amanita*

7. Stipe without bulbous base. Universal veil usually saccate, rarely friable .... Section *Vaginatae*

SUBGENUS *Amanita* SECTION *Amanita*

Here, and under a similar subheading for each section, we present a key to those taxa of the given section known from the region.

1. Pileus orangish red to brilliant red, fading on exposure to washed-out orange; universal veil yellow, at least at pileus margin when pileus begins to expand; basidiocarps medium to large; commonly under *Pinus*; spores with Q = 1.36–1.67 .... 1. *Amanita muscaria* var. *flavivolvata*

1. Pileus with yellow tones from early stages of expansion to maturity; universal veil white to off-white to dingy gray; basidiocarps small; under *Quercus*; spores with Q = (1–)1.02–1.19(–1.39) .... 2. *Amanita xylinivolva*

1. *Amanita muscaria* (L.:Fr.) Pers. var. *flavivolvata* (Sing.) Jenkins, *Biblio. Mycol.* 57: 56. 1977.

≡ *Amanita muscaria* subsp. *flavi-volvata* Sing., *Sydowia* 11: 374. 1958.

In Colombia, this taxon is known from plantations of imported *Pinus* spp. in Antioquia and Cundinamarca departamentos (Pulido-O., 1983; Henao-M. & Tobón-A., 1984). It is included here for completeness and to increase the usefulness of our keys in Colombia. A detailed description of this taxon is provided by Jenkins (1977, 1986). A collection made in Bogotá (*Plowman 2220-A*) is located in FH(!).

2. **Amanita xylinivolva** Tulloss, Ovrebo & Halling, *sp. nov.* Figs. 1–3, 29

Typus: COLOMBIA. ANTIOQUIA: Mpio. Guarne, near Centro Experimental Piedras Blancas, ±14 km E of Medellín, 11 Nov 1986, *Ovrebo 2487* (holotypus: HUA; isotypi: CSU, NY).

≡ [?]*Amanita gemmata sensu* Guzmán & Varela, *Caldasia* 12: 328. 1978, *non Amanita gemmata* (Fr.) Bertillon *in* DeChambre, *Dict. Encyc. Sci. Medic.* 1(3): 496. 1866.

Etymology: *xylinus*, cottony or pertaining to cotton + *volva*; describing the submembranous cottony universal veil material.

Pileus 15–60 mm latus, primo obscure flavus, dein cremeo-bubalinus; margo sulcatus vel tuberculato-striatus; materiae volvicae aut nullae aut assumenta albida vel sordido-grisea formantes. Stipes 45–100 × 4–9 mm, candidus; annulus initio ex frustris membranaceis constantes mox evanescentes; bulbus abruptus vel notatus, interdum subradicans; volva limbos submembranaceos gossypinos fugaces vertice bulbi formans. Fibulae nullae. Sporae inamyloideae, globosae vel subglobosae vel late ellipsoideae, aliquando ellipsoideae, (6.2–)8–10.2(–12) × (5.2–)7.2–9.5(–10.8) µm.

PILEUS: 15 – 60 mm in diam., hemispheric at first, expanding to broadly convex to plane, often becoming depressed to concave, slightly tacky, dull, when young dull yellow (duller than 4A4 – 5), lightening when older to dull creamy yellowish buff (including 3A2 – 3), sometimes remaining darker in disk, not discoloring; margin sulcate to tuberculate-sulcate with striations 0.25R – 0.5R, occasionally appendiculate with membranous bits of partial veil; context 2 – 4 mm thick, white to buff but yellow beneath pileipellis, unchanging when cut or bruised; universal veil absent or as white to off-white to

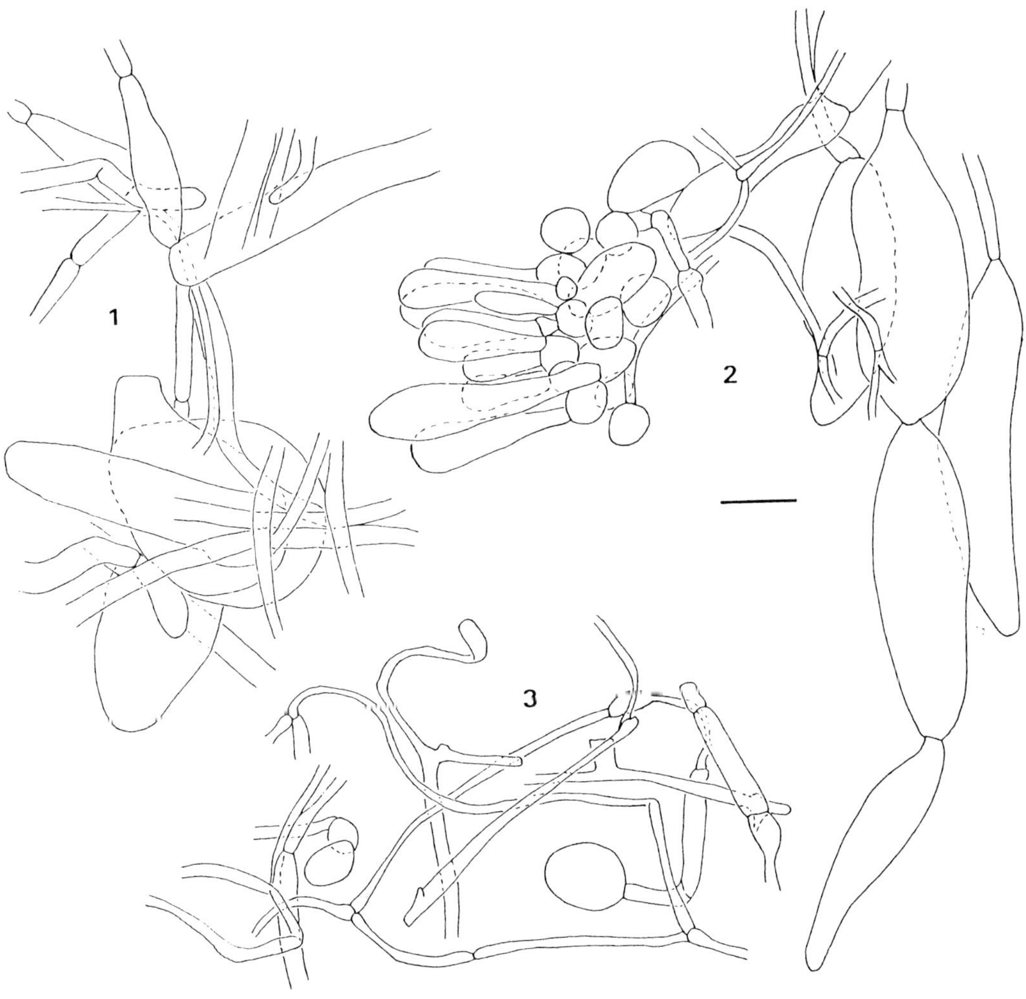

Figs. 1–3. *Amanita xylinivolva* (*Ovrebo 2487*). 1. Universal veil from pileus. 2. Hymenium and subhymenium. 3. Partial veil. Scale line = 20 µm.

dingy gray patches, scattered or concentrated over disk. LAMELLAE: free, subdistant to distant, whitish buff, not discoloring although somewhat dingy at pileus margin, drying to 5A–B4 to slightly grayer than 5B4, 2–8 mm broad, rounded at pileus margin; edges finely fimbriate (with 10× lens); lamellulae of considerably varying lengths, scattered, truncate. STIPE: 45–100 × 4–9 mm, white to creamy to whitish buff becoming dingy pale yellow, tapering slightly upward, finely fibrillose and shiny above, lower portion finely shaggy flocculent; context white to whitish buff, solid when young, slightly hollowed at maturity; annulus present as membranous flaps in young material, superior to subsuperior, soon slipping down on stipe and/or disappearing; bulb marked to abrupt, at times subradicating, woolly-fibrillose; universal veil as submembranous, cottony, fugacious flaps at top of bulb. Odor musty, earthy. Taste nondescript. Macrochemical spot tests: 3% KOH faint orange on pileus; 10% $FeSO_3$ negative. Spot test for laccase (syringaldazine) negative; spot test for tyrosinase (paracresol) positive throughout basidiocarp.

PILEIPELLIS: 75–150 μm thick, gelatinizing strongly at surface, orange brown in 10% NH_4OH; filamentous, undifferentiated hyphae 2–6.2 μm diam., branched, densely packed, somewhat interwoven, subradially arranged; oleiferous hyphae present, 2.8–5 μm diam. PILEUS CONTEXT: filamentous, undifferentiated hyphae 2–7 μm diam., interwoven, branched; inflated cells thin-walled, ovoid to clavate, to 66 × 40 μm; oleiferous hyphae uncommon, 1–2.5 μm diam. LAMELLA TRAMA: bilateral; filamentous, undifferentiated hyphae 1.8–7.5 μm diam., branched; inflated cells single or in short chains, to 98 × 35 μm, mostly 20–30 μm diam., dominating the trama, terminal, arising from hyphae in sometimes obscured central stratum, only last of a chain turning outward and then usually at an angle under 45°; oleiferous hyphae absent or uncommon and ±1.5 μm diam. SUBHYMENIUM: cellular to subcellular in a layer three to four cells thick, the cells sometimes in branching chains, subglobose to ellipsoid, to 22 × 13 μm, often with major axis roughly parallel to hymenial surface, larger cells farther from base of basidia; slightly inflated short hyphal segments present, but scarce in some regions; basidia arising singly or in pairs from inflated cells and short uninflated hyphal segments. BASIDIA: 36–62 × (9–)10–15 μm, 4-, 2-, and 1-sterigmate, with 4-sterigmate usually dominating but sometimes comprising only about 50% of basidia examined, thin-walled; clamps not observed. UNIVERSAL VEIL: On pileus: filamentous, undifferentiated hyphae 1.5–12.5 μm diam., branched, dominant in some regions, loosely interwoven near outer surface becoming more densely packed near gelatinization zone above pileipellis, terminal segments inflated to 18 μm diam.; inflated cells terminal, single or paired in a chain, thin-walled or occasionally with walls to 0.8 μm thick, plentiful, locally dominant, ovoid to elongate to narrowly clavate, to 71 × 46 μm; oleiferous hyphae uncommon to locally common or in tangles, 1.2–6.5 μm diam. Limb on top of bulb: similar structures, but with a larger proportion of filamentous, undifferentiated hyphae more tightly interwoven, gelatinizing. STIPE CONTEXT: acrophysalidic; filamentous, undifferentiated hyphae dominant, 2–11 μm diam.; acrophysalides thin walled, to 141 × 34 μm; oleiferous hyphae present, 1.5–9 μm diam. PARTIAL VEIL: a loose tangle of filamentous, undifferentiated, hyphae 2–8.5 μm diam., rather frequently branched, lacking any dominant orientation, dominant throughout much of the tissue, most of the narrower sort; small inflated cells scattered, with larger ones in clusters, terminal, subglobose to ellipsoid to ovoid (to 58 × 45 μm) or occasionally narrowly clavate (to 120 × 35 μm), with subterminal segments slightly inflated, rather short, and having constrictions at septa; oleiferous hyphae scarce or absent.

BASIDIOSPORES: [328/16/6] (6.2–)8–10.2(–12) × (5.2–)7.2–9.5(–10.8) μm [**L** = (8.5–)8.7–9.5(–9.8) μm; **L'** = 9 μm; **W** = 7.9–8.8(–9.3) μm; **W'** = 8.3 μm; Q = (1–)1.02–1.19 (–1.39); **Q** = 1.05–1.12(–1.15); **Q'** = 1.08], inamyloid, hyaline, thin-walled, smooth, globose to subglobose to broadly ellipsoid, occasionally ellipsoid, often slightly adaxially flattened; contents guttulate; apiculus sublateral to subapical, cylindric, often rather prominent; color in deposit not recorded.

Distribution and habitat: In April to May and November, gregarious under *Cupressus lusitanica* Franco and in *Quercus humboldtii* forest at 1800 to 2350 m elevation.

Collections examined: COLOMBIA. ANTIOQUIA: Mpio. Santa Rosa de Osos, Estación La Oculta, along road from San José de la Montaña to Aragón, 26 Nov 1986, *Ovrebo 2561* (HUA). CAUCA: Los Araoz, 12 Apr 1968, *Singer B6045* (F 1013493 as *A. frostiana*); Mpio. Popayán, vereda "Rio Blanco," Hacienda Cantaclaro (property of Mauricio Pinzón), 7 May 1990, *Franco-M. 421* (CUVC, NY). CUNDINAMARCA: Mpio. Pacho, between Pacho and Páramo Guerrero, near Pacho road, 9 Apr 1968, *Singer B6031* (F 1014392 as *A. gemmata*); Cerra de Las Mercedes, road between Bogotá and La Mesa, *s.d.*, *Guzmán 4627* (COL, ENCB *n.v.*). NARIÑO: Mpio. La Florida, ±27 km from Pasto, Hacienda El Barranco (property of Francisco Villareal), 11 May 1990, *Franco-M. 446* (NY, PSO).

DISCUSSION

Amanita xylinivolva (Fig. 29) is a small to medium sized, yellow to yellowish cream or buff-colored mushroom with sulcate to tuberculate-sulcate striations on the pileus margin. Cottony, pallid pieces of universal veil may be found on the pileus or on the point of juncture between the stipe and its marked bulb. This species falls within the ranges of habit and color that are currently interpreted in the Americas as *A. gemmata*, but its spores are subglobose.

This entity appears to have been collected and briefly described by Guzmán and Varela (1978) as *A. gemmata*. Among others noting the occurrence of "*A. gemmata*" in Colombia are Nasi (1977), Pulido-O. (1983), and Henao-M. and Tobón-A. (1984). There is a confusing cluster of taxa (some undescribed) to which North American mycologists currently apply the latter epithet, and a great deal of work remains to be done on this group. *Amanita xylinivolva* is clearly distinguishable from *A. gemmata* by having dominantly subglobose spores with Q' = 1.08. Jenkins (1986) stated that, for *A. gemmata s.l.* the Q' value is 1.39. European authors also describe ellipsoid spores for *A. gemmata* (*e.g.*, Gilbert, 1940-1941). For more about the Guzmán collection, see under "*A. inaurata* f. *americana*" *sensu* Sing., below under *A. sororcula*.

Species in sect. *Amanita* described with similar Q' and pallid to yellowish pilei include species that can be distinguished from *A. xylinivolva* in the following ways:

—*Amanita albocreata* (Atk.) E. J. Gilb. of the northeastern U.S. has a white to whitish pileus with yellow disk, is exannulate throughout development, has an ocreate bulb, and has slightly smaller spores (Tulloss & Jenkins, 1986; Jenkins, 1986).

—*Amanita crenulata* Peck of the northeastern U.S. has a pale beige or yellowish tan pileus that often has grayish tones, short marginal striations on the pileus, and universal veil remnants as pale, sordid beige pulverulence on the bulb and dense circles of small concentric warts on the pileus or absent. Tulloss (1990) measured over 1450 spores of this taxon because of its involvement in an ibotenic acid/muscimol-type poisoning. He found the range of Q to be (1.08–)1.11–1.25(–1.36); the spores of *A. crenulata* are less rounded than those of *A. xylinivolva*.

—*Amanita diemii* Sing., recorded from Argentina and Chile (Singer & Moser, 1965), has an exannulate stipe which is much thicker in proportion to its length than is the stipe of *A. xylinivolva* and a bulb reminiscent of that seen in *A. pantherina* (DC.:Fr.) Krombh. (Garrido & Bresinsky, 1985). Singer (1954) reported that the stature is such that, after a hard rain, *A. diemii* can be taken for a *Russula*. The pileus of *A. diemii* often becomes brownish in age and may even become quite a dark brown (Singer, 1954); this contrasts to that of *A. xylinivolva* which becomes pallid with age. In the protologue of *A. diemii*, Singer (1954) described spores very much like those we found in *A. xylinivolva*. Chilean material determined as *A. diemii* by Singer and Moser (1965) had spores with Q' of about 1.2 and of dimensions falling almost completely outside the range found for *A. xylinivolva*.

—*Amanita elata* (Mass.) Corner & Bas (1962), described from Singapore, exhibits some similarities to *A. xylinivolva* in habit, shape of spores, and anatomy of the lamella trama and subhymenium. However, the spores of *A. elata* are notably smaller [7–8.5 × (6–)6.8–7.7 μm] than in *A. xylinivolva*. *Amanita elata* has an unpleasant odor as in *A. phalloides* (Fr.) Link and a pileus that is more or less umber or fuligineous on the disc. It is exannulate throughout development.

—*Amanita levistriata* Jenkins, known from the U.S. coast of the Gulf of Mexico, has a persistent, yellowish partial veil, an universal veil that is left as flocculence on the bulb rather than in submembranous fragments, a pileus margin that is only lightly striate, and slightly smaller spores (Jenkins, 1988). Tulloss has examined fresh material and spores of several of the paratypes of *A. levistriata* (in RET), and his findings support those of Jenkins.

—*Amanita xerocybe* Bas from Amazonia has spores with Q = 1–1.05, a pileus that darkens to brownish tones with age rather than becoming lighter, an annulus that is more persistent, and a more floccose universal veil (Bas, 1978).

We also considered comparison to *A. silvatica* Guzmán (1982) described from Mexico; however, this entity is smaller with yellowish tints in the universal veil, much shorter basidia, narrower spores, etc. It is possible that *A. xylinivolva* includes other Mexican material placed by various authors in *A. gemmata* or *A. gemmata* f. *gracilis sensu* Santiago *et al.* (1984).

SUBGENUS *Amanita* SECTION *Vaginatae* (Fr.) Quél.

1. Universal veil not membranous (pulverulent to submembranous).

 2. Universal veil fulvous at first, becoming dark reddish to fuscous brown; spores with Q = 1–1.07(–1.17) .... 3. *Amanita colombiana*

 2. Universal veil pallid or whitish at first, becoming grayish; spores with Q = (1–)1.01–1.16(–1.25) .... 6. *Amanita sororcula*

1. Universal veil membranous.

 3. Subhymenium 1 to 3 cells deep; spores (7.5–)9–12(–15.5) × (6.5–)8.2–11.2(–15) μm, with L = (9.5–)9.6–11.4(–11.5) μm, with W = (8.6–)8.9–10.6(–10.7) μm, and with Q = (1–)1.02–1.18(–1.4) .... 4. *Amanita fuligineodisca*

 3. Subhymenium 4 to 5 cells deep; spores (10.2–)10.8–14(–15.5) × (9–)9.7–13(–13.3) μm, with L = 12.5–12.8 μm, with W = 11.4–11.8 μm, and with Q = 1–1.17(–1.28) .... 5. *Amanita humboldtii*

3. **Amanita colombiana** Tulloss, Ovrebo & Halling, *sp. nov.* Figs. 4–5, 31

Typus: COLOMBIA. ANTIOQUIA: Mpio. Santa Rosa de Osos, vereda "El Chaquiro," road from Santa Rosa de Osos to Aragón, near Llanos de Cuivá, 5 Nov 1986, *Ovrebo 2425* (holotypus: HUA).

Etymology: Named for the Republic of Colombia.

Pileus 22–55 mm latus, obscure olivaceo-brunneo-griseus; margo plicato-sulcatus; materiae volvicae e verrucis pyramidalibus vel conicis interdum ex assumentis complanatis constantes, initio fulvae demum fuscescentes vel atro-rubro-brunneae vel fusco-brunneae. Stipes 75–80 × 5–6 mm, sordido-albus, exannulatus; assumenta volvae fulva gossypina annulum completum vel incompletum formantia stipitis basi adpressa. Fibulae nullae. Sporae inamyloideae, globosae vel subglobosae, (9–)10–12.2(–14.2) × (8.5–)9.5–12(–14) μm.

PILEUS: 22–55 mm diam., tacky, subpolished, smooth, dull olive brownish gray (near 81. d. gy. y Br), not discoloring; margin nonappendiculate, plicate-sulcate (striations 0.4R–0.75R); context 3–4 mm thick, white; universal veil as pyramidal to conical warts, occasionally flattened on top, scattered or often clustered near center, fulvous at first (between 51. deep O and 55. s. Br), darkening in age to dark reddish brown or fuscous brown. LAMELLAE: free, subdistant at maturity, white, not discoloring, drying close to 4A4 to a little more cream than 4A4 to 5B4, 4–6 mm broad, rarely forking near pileus margin; edge grayish marginate when mature, very finely fimbriate (with 10× lens); lamellulae numerous, not in regular tiers, truncate. STIPE: 75–80 × 5–6 mm, dingy white overall, not discoloring, tapering slightly to apex, very finely and sparsely appressed fibrillose with fibrils largest and coarsest near base; context white, hollow; exannulate; universal veil in patches of fulvous, cottony tissue, forming incomplete to complete ring above a similar portion appressed to very base of stipe, woolly, and fulvous (7–8D8). Odor and taste absent. Macrochemical spot tests: 2.5% KOH on pileus and lamellae, negative.

PILEIPELLIS: 40–100 μm thick, gelatinizing at the surface, hyaline at least in the upper half, palely tinted with pigmentation of context below; filamentous, undifferentiated hyphae 2–10.2 μm diam., infrequently branched, interwoven, subradially arranged; oleiferous hyphae present, 2.8–10 μm diam., rarely branched, locally common. PILEUS CONTEXT: filamentous, undifferentiated hyphae 2.8–14.5 μm diam., interwoven, branched, hyaline or with orange-brown to brownish intracellular pigment sometimes forming reddish brown particles or guttules; inflated cells ellipsoid to elongate to subclavate, to 123 × 52 μm; oleiferous hyphae plentiful, 2–10.5 μm diam., branched, locally in tangles. LAMELLA TRAMA: bilateral; filamentous, undifferentiated hyphae 2.8–19.5 μm diam., branched, some segments slightly inflated; inflated cells to 86 × 52 μm; oleiferous hyphae present, 1–2.5 μm diam. SUBHYMENIUM: subcellular, ramified; basidia arising from short, branched, uninflated to slightly inflated hyphal segments, also arising occasionally from inflated cells 8–11.5 μm diam. BASIDIA: 43–81 × 13.2–19.8 μm, 4-, or occasionally 2-sterigmate, thin-walled; sterigmata rather broad at the base, to 2.2 μm diam.; clamps not observed. UNIVERSAL VEIL: On pileus: all elements somewhat gelatinizing, dissociated; filamentous, undifferentiated hyphae 2.3–6.4 μm diam., branched, infrequent, fragmented; inflated cells numerous and dominant, subglobose to ellipsoid to ovoid to clavate, to 57 × 34 μm, thin- to slightly thick-walled, collapsed, dull brown when inflated, but losing pigment upon collapsing, orange-brown in mass in exsiccata; oleiferous hyphae uncommon, fragmented, 2–2.5 μm diam. On stipe base: branched, filamentous, undifferentiated hyphae 3–9.5 μm diam., more plentiful than on pileus; inflated cells with brown intracellular pigment, dominant, globose to subglobose to ellipsoid to ovoid to clavate as on pileus, to 67 × 49 μm, terminal or occasionally in short chains, with walls 0.5–1.5 μm thick, orange-brown in mass in exsiccata; oleiferous hyphae scarce or absent. STIPE CONTEXT: acrophysalidic; undifferentiated hyphae 1.2–13.2(–21) μm diam., dominant, branched, filamentous, with ochraceous tint near surface; acrophysalides to 278 × 70 μm, mostly less than half that size, with thin walls or walls to 0.8 μm thick; branched, oleiferous hyphae 1.5–12.5 μm diam., common to plentiful, locally in knots.

BASIDIOSPORES: [106/4/3] (9–)10–12.2(–14.2) × (8.5–)9.5–12(–14) μm [**L** = 10.8–11.5 μm; **L'** = 11.3 μm; **W** = 10.5–11.1 μm; **W'** = 10.9 μm; Q = 1–1.07(–1.17); **Q** = 1.03–1.04; **Q'** = 1.03], inamyloid, hyaline, thin-walled, smooth, globose to subglobose; contents granular to guttulate; apiculus sublateral, cylindric; color in deposit not recorded.

Distribution and habitat: Solitary or in pairs, in November, under *Quercus humboldtii*, at ±2500 m elevation.

Collections examined: COLOMBIA. ANTIOQUIA: Mpio. San José de la Montaña, Estación La Oculta, along road from San José de la Montaña to Labores, 25 Nov 1986, *Ovrebo 2541* (HUA), 26 Nov 1986, *Ovrebo 2560* (HUA).

DISCUSSION

Amanita colombiana (Fig. 31), a small to medium sized mushroom, is most remarkable for its collapsing, friable to submembranous universal veil that becomes distinctly reddish as the basidiocarp expands. The pileus of *A. colombiana* is a dull brownish gray with rather marked marginal striations that may be as long as three-quarters of the pileus radius. The stipe is exannulate. The unusual universal veil is left as patches on the pileus and as a broken cup-like volva at the stipe base.

The color of the universal veil in *A. colombiana* is unique among the species described in section *Vaginatae*. Red spots are said to appear on the universal veil of *A. constricta* Thiers & Ammirati (1982) in moist conditions, and that species (described from northern California) does have plentiful oleiferous hyphae in the pileus context. *Amanita constricta*, however, has a more nearly submembranous universal veil that is mostly gray and spores with Q ≈ 1.2.

4. **Amanita fuligineodisca** Tulloss, Ovrebo & Halling, *sp. nov.* Figs. 6–7, 30

Typus: COLOMBIA. ANTIOQUIA: Mpio. Santa Rosa de Osos, vereda "El Chaquiro," road from Santa Rosa de Osos to Aragón, near Llanos de Cuivá, 14 Nov 1986, *Halling 5022* (holotypus: HUA; isotypus: NY).

Etymology: *fuligineus*, a brown which is nearly black, + *discus*, disc.

Pileus 27–70+ mm latus, disco atrobrunneo vel atrofulvo vel fere nigro; margo interdum pallidior, plicato-striatus; materiae volvicae plerumque nullae vel interdum assumentum membranaceum formans. Stipes 70–150 × 3–12 mm, fulvus vel super cremeus ut infra albus vel omnino albus, contusus fulvescens vel grisescens, exannulatus; volva saccata, alba vel bubalina vel pallide aurantio-alba. Fibulae nullae. Sporae inamyloideae, globosae vel subglobosae, (7.5–)9–12(–15.5) × (6.5–)8.2–11.2(–15) µm.

PILEUS: 23–70+ mm diam., dark brown (including 5F8 to 6F4–5 to 6F7) to dark fulvous (including 6E8) or nearly black at disc (including 8F4), at times paler (including 5B2–4 to 5D3 to 6C4–5 to 6D4 to 5F4) or fulvous toward margin or (infrequently) overall, convex to plane to depressed in age, with slight umbo, viscid when moist; margin plicate-striate (0.25R–0.65R), nonappendiculate, at times becoming rimose; context white, unchanging when cut, 2–4+ mm thick; universal veil almost always absent, but occasionally present as membranous patches concolorous with saccate volva. LAMELLAE: free to very narrowly adnate, close to subdistant, white to buff, becoming sordid tan to sordid brown (4A4 to 5–7.5YR 4/6 to 7.5YR 6/6 to 6D4) on drying, 4.5–7 mm broad, not or rarely marginate and then having an edge concolorous with pileus; lamellulae truncate, of varying lengths. STIPE: 70–150+ × 3–12 mm, fulvous or creamy above and white below or entirely white or whitish with fulvous, orange, or grayish tones after handling (including near 5A–B2–3 to a little darker than 5A2), tapering upward, fibrillose (with 10× lens) or with somewhat darkened (brownish) fibrils; exannulate; context becoming hollow; universal veil saccate, white to buff to pale orangish white (5–6A2), often with orange-fulvous to ochraceous brown to rather dark brown stains, adpressed at the base of stipe (as in *A. fulva*), free above. Odor and taste not recorded. Macrochemical spot tests: negative for laccase; positive throughout basidiocarp or only in stipe context for tyrosinase.

PILEIPELLIS: surface layer completely gelatinizing and 5–10 µm thick; below this a layer 10–30 µm thick of tightly interwoven, branched, filamentous, undifferentiated hyphae 1.8–6.8 µm diam., without a dominant orientation; oleiferous hyphae 1.2–6.5 µm diam., branched, locally common. PILEUS CONTEXT: filamentous, undifferentiated hyphae 2–13.2 µm diam., branched; inflated cells thin-walled, terminal, broadly ovoid to ovoid to pyriform to 90 × 72 µm, elongate to 182 × 39 µm; oleiferous hyphae 1.5–5 µm diam. LAMELLA TRAMA: bilateral, at times obscurely so, dominated by interwoven hyphae of the central stratum and similar hyphae running parallel to it against which the relatively thin subhymenium is pressed; undifferentiated hyphae 1–13 μm diam., branched, filamentous; inflated cells to 56 × 36 μm mostly parallel to central stratum or at an angle of 30° or less to the central stratum, some having walls slightly thickened (rarely to 0.8 μm thick); oleiferous hyphae 1.5–9 μm diam., branched, occasionally locally in knots. SUBHYMENIUM: a thin layer one to three cells deep of globose, small cells, mostly under 12 μm diam., and ellipsoid to ovoid cells to 25 × 15 μm; basidia arising from these cells or from the sides (not the tips) of slightly inflated terminal segments of hyphae running roughly parallel to the central stratum. BASIDIA: 37–73 × 11.8–19.5 μm, dominantly 4-sterigmate, occasionally 2-, rarely 1- or 3-spored, thin-walled; sterigmata quite broad (to 2.5 μm) at base; clamps not observed. UNIVERSAL VEIL: On stipe base, external surface: filamentous, undifferentiated hyphae 3.8–14 μm diam., loosely interwoven, slightly gelatinized, branched, without obvious orientation, or at times in gelatinized fascicles; very few inflated cells, *e.g.*, 50 × 39 μm; oleiferous hyphae 2.6–6.4 μm diam. On stipe base, interior: a loose tangle of branched, filamentous, undifferentiated hyphae dominant, 2.6–14.3 μm diam.; inflated cells plentiful (more so at a greater distance from either surface), thin-walled, subglobose to ovoid or ellipsoid to 81 × 60 μm, clavate to 62 × 27 μm; oleiferous hyphae ±3 μm diam. On stipe base, inner surface: slightly gelatinized, otherwise very like the interior but with greater dominance of more tightly interwoven filamentous, undifferentiated hyphae; oleiferous hyphae ±4 μm diam. On pileus: absent. STIPE CONTEXT: acrophysalidic; filamentous, undifferentiated hyphae 2–7 μm diam., branched; acrophysalides to 200 × 38 μm, thin-walled; oleiferous hyphae ±6.5 μm diam., uncommon.

BASIDIOSPORES: [585/29/12] (7.5–)9–12(–15.5) × (6.5–)8.2–11.2(–15) μm [**L** = (9.5–)9.6–11.4(–11.5) μm; **L'** = 10.6 μm; **W** = (8.6–)8.9–10.6(–10.7) μm; **W'** = 9.9 μm; Q = (1–)1.02–1.18(–1.40); **Q** = (1.05–)1.06–1.12(–1.15); **Q'** = 1.08], inamyloid, walls thin or occasionally slightly thickened as seen in Melzer's reagent, smooth, hyaline, globose to subglobose, sometimes slightly adaxially flattened, occasionally with adaxial indentation; contents guttulate; apiculus prominent and rather large (to 2.5 μm diam. and sometimes projecting to a length greater than the diameter), cylindric or rarely truncate conic, subapical to sublateral; color in deposit not recorded.

Distribution and habitat: Under *Quercus humboldtii* at 2500–2750 m (or higher) elevation, May to June and November.

Collections examined: COLOMBIA. ANTIOQUIA: Mpio. Santa Rosa de Osos, vereda "El Chaquiro," on the road from Santa Rosa de Osos to Aragón, near Llanos de Cuivá, 10 Nov 1986, *Ovrebo 2482* (HUA, NY, CSU), 13 May 1987, *Halling, Mueller & Strack* [*GMM no. 2818*] (F 1080753), 8 Nov 1988, *Franco-M. 104* (HUA, NY), 12 Jun 1990, *Franco-M. 521* (HUA, NY), 22 Jun 1990, *Franco-M. 536, 538* (both HUA, NY). BOYACÁ: Mpio. San Miguel de Sema, road from Simijaca to San Miguel de Sema, 9 May 1987, *Halling 5246* (COL, NY). NARIÑO: Mpio. La Florida, ±27 km from Pasto, Hacienda El Barranco (prop. F. Villareal), 11 May 1990, *Franco-M. 444* (NY, PSO); Mpio. Pasto, 8 km E of Chachagüí, bosque "El Común," 22 May 1990, *Franco-M. 465* (NY, PSO), vereda "La Josefina," km 17, road from Pasto to Chachagüí, 22 Nov 1988, *Franco-M. 193* (PSO), 9 May 1990, *Franco-M. 427* (NY, PSO).

Discussion

Amanita fuligineodisca (Fig. 30) has a dark brown to blackish brown pileus which is sometimes paler toward the striate margin. It suggests *A. fulva* (Schaeff.) Pers. of more northern countries (originally described from Bavaria) in its habit and in its whitish, saccate volva which often becomes stained rusty and is adpressed to the stipe in its lower portion. The stipe is exannulate. Pileus color, lamella trama, subhymenium, and spore size differentiate this entity from *A. fulva*.

Since almost all the collections of *A. fuligineodisca* examined were considered in the field to belong possibly to *A. fulva*, a detailed comparison is appropriate. Moreover, one of the reviewers of this paper (Dr. C. Bas, Leiden) raised significant qustions concerning the taxonomic relationship of the two entities; these concerns deserve a considered response.

Amanita fulva is treated as common in contemporary European literature, in much of eastern North America (Jenkins, 1986), and is reported from Japan (Imazeki & Hongo, 1987) and from sub-arctic to arctic locations such as Greenland (Knudsen & Borgen, 1987). Apparently, *A. fulva* is widespread in the Northern Hemisphere. To our knowledge, it has not been reported from South America.

The spores of *A. fulva* are generally reported to be nearly perfectly globose; for example, Jenkins (1986) gave the value of **Q'** for *A. fulva* as 1.02. In 1988, Tulloss had the opportunity to collect a number of specimens of *A. fulva* in the United Kingdom—northeastern Scotland, Cumbria, and Dorset (all in RET). Although the spores from these collections of *A. fulva* (collection numbers: *9-5-88-A*, *9-5-88-G*, *9-5-88-O*, *9-10-88-B*, *9-13-88-B*) are similar to those seen in *A. fuligineodisca*. The spores of *A. fulva* are slightly larger on average and are somewhat more nearly globose.

BASIDIOSPORES of U.K. collections of *A. fulva*: [100/5/5] (9–)9.8–12.2(–15) × (8.2–)9.5–11.8(–13.8) μm [**L** = 10.6–11.8 μm; **L'** = 11.1 μm; **W** = 10.1–11.1 μm; **W'** = 10.5 μm; Q = 1–1.11(–1.21); **Q** = 1.05–1.06; **Q'** = 1.06].

Dr. C. Bas obtained material of *A. fulva* from Schaeffer's original collecting region (Bavaria). In this material (18 Oct 1987, *N. Arnold s.n.*, L boxes I–IV), the spore characters are the same as those seen in the U.K. collections.

BASIDIOSPORES of Bavarian specimens of *A. fulva*: [40/2/1] (9–)9.3–13.6(–15) × (8.4–)8.8–13(–14.4) μm [**L** = 11–12 μm; **L'** = 11.5 μm; **W** = 10.3–11.4 μm; **W'** = 10.8 μm; Q = 1–1.11(–1.21); **Q** = 1.05–1.07; **Q'** = 1.06].

The differences in size and shape of the spores of *A. fulva* and *A. fuligineodisca* are statistically significant ($p<.01$ in 'F' test). Nevertheless, we do not base our argument separating the taxa on differences in the size and shape of the spores.

The very dark disk (even entire pileus) of *A. fuligineodisca* is not a character seen in *A. fulva*; and the lamella trama of *A. fulva* contains more abundant inflated cells (to 76 × 44 μm) even to the extent of obscuring the central stratum (*e.g.*, in *Tulloss 9-5-88-F*, RET). Moreover, these cells diverge at an angle of 60–70° from the central stratum although sometimes with an intervening cell at a shallower angle. The angle of divergence of the trama of *A. fuligineodisca* is under 30°. In this species, basidia arise from inflated cells and the sides of segments of undifferentiated hyphae; whereas in *A. fulva*, basidia arise from inflated cells and the tips of uninflated or slightly inflated hyphal segments.

We believe these are sufficient differences to establish *A. fuligineodisca* as a distinct species.

⟶

Figs. 4–12. Habit and microscopic features. Figs. 4–5. *Amanita colombiana* (*Ovrebo 2425*). 4. Hymenium and subhymenium. 5. Dissociated elements of universal veil from pileus. Figs. 6–7. *Amanita fuligineodisca* (*Halling 5022*). 6. Hymenium and subhymenium. 7. Universal veil from stipe base. Figs. 8–10. *Amanita humboldtii* (*Singer B3527*). 8. Habit per field sketch of R. Singer. 9. Hymenium and subhymenium. 10. Universal veil from stipe base. Figs. 11–12. *Amanita sororcula* (*Ovrebo 2507*). 11. Hymenium and subhymenium. 12. Universal veil from stipe base, near exterior surface. Stippling indicates refractive content. Scale line = 10 μm (10 mm in Fig. 8).

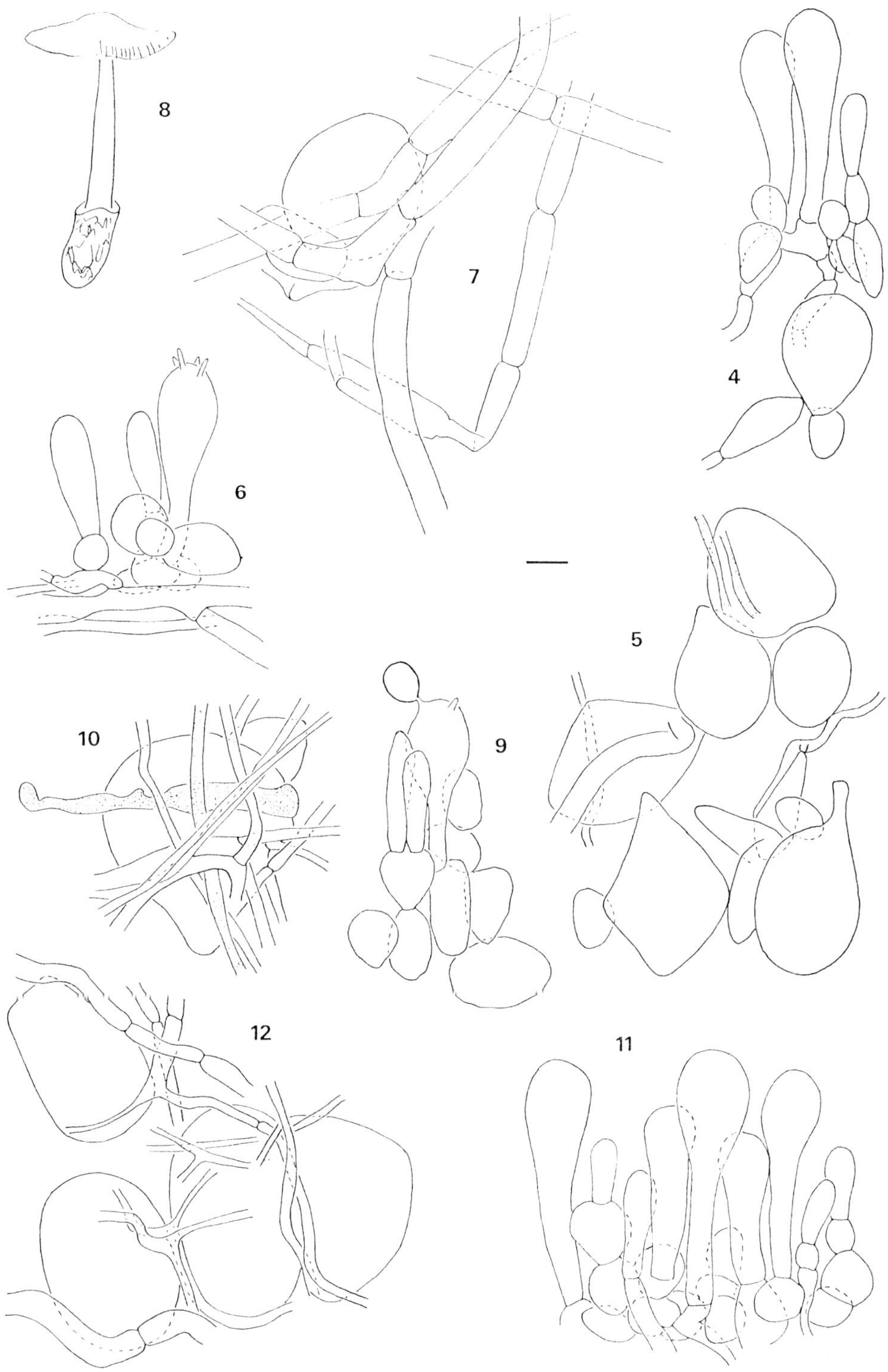
8
7
4
6
5
10
9
12
11

Elsewhere in this paper, we have taken the position that with regard to *Amanita*, differences limited to pileus color and spore size/shape (albeit statistically significant) do not justify introduction of an entity at the level of species. The additional differences in this case relate to anatomy of lamella trama and subhymenium. Appeals to these particular anatomical details in discussion of *Amanita* taxonomy (especially in the erection of new species) is at least uncommon. In particular, we must address the question: Is it not the case that the cool, wet environment of Scotland may have induced a greater degree of inflation of cells of the lamellae in *A. fulva*, consequently altering the arrangement from that found in *A. fuligineodisca* and obscuring the fact that all the relevant material is conspecific?

That spore size and shape may be altered by the amount of moisture in the environment is commonly asserted. Tulloss has done an extensive study of distribution vs. spore size and shape in *A. peckiana* Kauffman and other taxa of *Amanita* section *Amidella* (unpublished data). The data collected provide strong evidence that spores of *A. peckiana* from wet, mountainous, western North Carolina are broader, shorter, and of lower Q than spores of the same species from dry, hot, sandy, eastern Texas. But does this entail a significant alteration in the structures supporting the basidia? It is characteristic of the holotype of *A. peckiana* that its subhymenium and lamella trama contain plentiful hyphae barely showing divergence from the central stratum; this character is to be seen in the material from both very wet and very dry regions.

Amanita morrisii Peck (section *Validae*)is typified by a specimen collected in a swamp. Tulloss has reviewed abundant material from holotype, isotype, paratype, and topotype collections; this species was always found to have a subhymenium containing plentiful, uninflated, nondiverging hyphae (Tulloss, 1991). Tulloss (1990) examined a large number of specimens of *A. crenulata* Peck (section *Amanita*), and in this species, the lamella trama did not vary markedly even though some specimens were from low, wet areas (in mixed woods of *Acer, Betula,* and *Liquidambar styraciflua* L.) and others from dry, sandy, permeable soils of the Atlantic Coastal plain, and pine-oak barrens.

In an as yet unpublished study of white taxa with limbate volvas in *Amanita* section *Phalloideae* in North America, Tulloss has examined collections of *A. magnivelaris* Peck collected in very moist, mountain habitat. When fresh, the basal bulbs of these specimens were swollen and as firm as a raw potato. However, the subhymenium and lamella trama were easily recognizable as the same in form and arrangment as those elements observed in slenderer, soft-bulbed collections from drier terrain.

In summary, we rely on the cited studies. These do not demonstrate any case of environmentally induced cell inflation in the tissues under discussion—certainly not to the degree necessary to suggest that *A. fuligineodisca* would have the lamella anatomy of *A. fulva* were it to be collected in a wet environment.

Section *Vaginatae* is not in a well understood or well ordered state. We have given careful consideration to the recognition of this particular new species because there are approximately forty possibly distinct taxa in the section described in the literature as having a membranous universal veil in the form of a saccate volva and globose to subglobose spores. If we eliminate those that are normally of very large stature, or have an extremely thick universal veil, or are not brown, or have a zonate pileus, or have spores of a markedly different size than those of *A. fuligineodisca*, then the problem is reduced. There is still difficulty due to the absence of type specimens for some of the older European names. Taxa phenetically similar to *A. fuligineodisca* (other than *A. fulva* are as follows:

–*Amanita aporema* Boedijn (1951) was described from Indonesia; its pileus is red-brown and tends to be rather large (to 140 mm diam.). Its stipe is dirty white, not showing the tints described for *A. fuligineodisca*; and its volval sac can be relatively long (40–60 mm). No staining of the universal veil is mentioned.

—*Amanita humboldtii* is described in detail below. It differs from *A. fuligineodisca* in its spore size and in the structure of the subhymenium and lamella trama. Even though the ranges of spore length and width for the two species overlap, the mean lengths and mean widths are significantly different ($p < .01$ for 'F' test). We have not yet had the opportunity to compare the species macroscopically in the field, and to our knowledge, no color photographs of *A. humboldtii* have been made. The two species must be rather similar macroscopically-with the pileus of *A. humboldtii* being smoky brown while that of *A. fuligineodisca* is blackish brown in the disk and often has fulvous tones near the margin. *Amanita fuligineodisca* appears to be the more common of the two in the areas in which collections reviewed for this paper have been made.

—*Amanita sampajensis* Sathe & Kulkarni (Sathe *et al.*, 1980), with an isabelline to hazel colored cap is known from southwestern India. It is described as having a gelatinous layer 35 – 75 μm deep on the pileal surface in which are found narrow, wavy hyphae with olivaceous contents; its universal veil is not reported as staining and is described as reaching a height of 62 mm; its spores are somewhat larger than those of *A. fuligineodisca*: 10.5 – 13.5 × 9.5 – 12 μm.

—*Agaricus badius* Schaeff. (1774) poses the greatest problem for us. There is no type specimen, and it has been either interpreted diversely by twentieth century European authors or ignored by them. Gilbert (1918) summarized European views on Schaeffer's entity up to that time and treated it as a variety of *Amanita vaginata* (Bull.:Fr.) Vitt. Gilbert (1940-1941) considered it to have a "*cutis umbrino-castanea*," a volva "*pallide castaneo-fulva*," and spores 11 – 13 μm diam. Further, Gilbert said that it is rare and that he had never seen it. We note that *A. fuligineodisca* appears to be one of the most commonly collected species of *Amanita* in our study area of Colombia. Quélet (1888) mentioned Schaeffer's entity as a form of *A. vaginata* that is "*bai brun ou bistre.*" Heinemann (1964) wrote only that the pileus is chestnut brown and darker than that in *fulva*. Parrot (1960), Marchand (1971, 1973), Moser (1983), and Kühner and Romagnesi (1974) did not mention it. Merlo and Traverso (1983) provided an illustration to which Schaeffer's name is applied, but gave no additional information. We know of only three references in North American literature other than in regional collecting lists. Peck (1879) treated Schaeffer's entity as a mere color variant of *A. vaginata*. Atkinson (1914) simply cited the name and Schaeffer's original plate number (245 in *Icones Fungorum*). Krieger's (1935) interpretation was idiosyncratic, possibly related to an undescribed North American entity, and sounds surprisingly similar to *A. fuligineodisca* except for the claim of rarity and the association with swamps: "very dark-brown, almost black in the center (rare and found only in swampy ground of mountainous regions)." At the very least, Krieger's entity seems to have different coloration than in the European interpretations. If *Agaricus badius* is distinct from *Amanita fulva*, the literature indicates the former to be closely related to the latter and thus to be likewise distinct from *A. fuligineodisca*. The situation will not be satisfactorily resolved until a "representative specimen" (Mueller, 1991:665) is designated for *Agaricus badius* and its anatomy carefully described.

Franco-M. 427, 465 (NY, PSO), collected in Nariño, are included here because they were not distinguishable anatomically from the holotype. However, the field descriptions and color transparencies accompanying the collections show a white, unchanging, saccate volva as well as a white stipe without orange tints and bearing scattered brown fibrils. Also, a paracresol spot test on *Franco-M. 427* reacted positively for tyrosinase only in the stipe. Without support at the anatomical level for taxonomic differentiation, the collections in question are interpreted as evidence of variation within the species.

Halling 6160 (HUA, NY), collected in Antioquia, appears to be assignable to this species; however, no mature spores were found on the lamellae.

5. *Amanita humboldtii* Singer, *Mycopathol. Mycol. Appl.* 20: 241. 1963. Figs. 8–10

Typus: COLOMBIA. CUNDINAMARCA: Mpio. Pacho, Los Robledales [misspelled in protologue] above Pacho, 26 Jul 1960, *Singer B3527* (holotypus: BAFC 31.975).

PILEUS: 37–60 mm diam., "cochin" (MP) (dull fuscous brown) or 6D–F8 in disk and 6C7 or 6E6 toward margin, plano-convex, subumbonate to depressed, slightly viscid when wet; margin pectinate-sulcate, decurved; context white, unchanging on exposure, ±3.5 mm thick; remnants of universal veil absent. LAMELLAE: free, close, white, drying a pale slightly sordid tan or brown (6D–E4), ±5 mm broad, with fimbriate edge that may be white or concolorous with pileus surface near pileus margin. STIPE: 63–180 × 6–11 mm, firm, narrowing upward, entirely white or somewhat darker than 5B3 in upper half and white elsewhere, glabrous to fibrillose; exannulate; context white, solid or becoming hollow especially in the lower portion; universal veil as a "caramel" (MP) (cinnamon to orange) or white, membranous, free volval sac, *e.g.*, 20 × 12 mm. Odorless. Taste not recorded.

PILEIPELLIS: 20–56 μm thick; filamentous, undifferentiated hyphae 1.2–7.8 μm diam., slightly gelatinized, subradially arranged, interwoven, with a pale yellowish brown intracellular pigment partially soluble in KOH; oleiferous hyphae 1.8–11 μm diam., branched, relatively common. PILEUS CONTEXT: filamentous, undifferentiated hyphae 1.2–12.8 μm diam., loosely interwoven to tangled, sometimes in fascicles, with some intercalary segments slightly inflated to 14 μm diam.; inflated cells with walls thin or to 0.5 μm thick, slenderly clavate to subventricose to ventricose (to 92 × 27 μm) to broadly clavate (to 71 × 42 μm) to ellipsoid to subglobose or pyriform (to 63 × 51 μm); branched oleiferous hyphae common, 1.5–10.5 μm diam. LAMELLA TRAMA: bilateral; with a very distinct central stratum; angle of divergence 45°–90°; filamentous, undifferentiated hyphae 2.1–11.5 μm diam., branched; inflated cells to 34 × 20 μm, clavate to ventricose to ellipsoid, diverging at angles cited above; oleiferous hyphae 2.2–6.3 μm diam. SUBHYMENIUM: 4 to 5 inflated cells deep; basidia arising from uninflated to slightly inflated hyphal segments or (predominantly) from inflated cells of varying size (to 27 × 25 μm, clavate, broadly clavate, ovoid, ellipsoid, subpyriform, subglobose, globose, arranged in branched chains arising in the central stratum). BASIDIA: 49–78 × 11.5–24 μm, thin-walled, clavate, dominantly 4-, occasionally 2-sterigmate; sterigmata to 2 μm broad at base; clamps rare. UNIVERSAL VEIL: At stipe base, exterior surface: a very thin layer of gelatinized, filamentous, undifferentiated hyphae 2.2–12.6 μm diam., dominant, arranged in a disorderly, loose mat, also occasionally in longitudinally oriented fascicles; scarce inflated cells to 39 × 33 μm, gelatinized; oleiferous hyphae scarce or absent. At stipe base, interior: filamentous, undifferentiated hyphae 1.4–15.8 μm diam., rather densely tangled/interwoven, plentiful, branched, locally dominant, at times in fascicles, somewhat gelatinizing; inflated cells more plentiful than at surfaces, locally dominant, subpyriform to ellipsoid to ovoid to broadly clavate to ellipsoid, thin-walled, to 65 × 54 μm; oleiferous hyphae 2.8–7 μm diam. On stipe base, inner surface: a rather thin layer at times absent in some specimens or in isolated areas of the volval limb of other specimens; filamentous, undifferentiated hyphae 1.5–8 μm diam., branched, longitudinally arranged, partially gelatinized; inflated cells scattered, collapsing, gelatinized, to 52 × 34 μm; oleiferous hyphae 4.2–17 μm diam., partially gelatinized, sublongitudinally arranged to disordered. On pileus: absent. STIPE CONTEXT: acro-physalidic; filamentous, undifferentiated hyphae 2–10.5 μm diam.; acrophysalides narrowly clavate to narrowly subventricose, to 290 × 39 μm, with walls thin or to 0.8 μm thick; oleiferous hyphae 1.2–13.3 μm diam., occasionally tangled in knots.

BASIDIOSPORES: [67/2/2] (10.2–)10.8–14(–15.5) × (9–)9.7–13(–13.3) μm [**L** = 12.5–12.8 μm; **L'** = 12.7 μm; **W** = 11.4–11.8 μm; **W'** = 11.7 μm; Q = 1–1.17(–1.28); **Q** = 1.08–1.09; **Q'** = 1.09], inamyloid, hyaline, smooth, globose to subglobose to broadly ellipsoid; contents guttulate to granular, in Melzer's reagent coagulating and adhering to spore walls; apiculus sublateral, often broad and prominent, cylindrical; color in deposit not recorded.

Distribution and habitat: In *Quercus humboldtii* forests in July and November, solitary, at 1700–2700 m elevation.

Collection examined: COLOMBIA. NARIÑO: Mpio. Pasto, vereda "La Josefina," km 17, road from Pasto to Chachagüí, 20 Nov 1988, *Franco-M. 190* (NY, PSO).

DISCUSSION

Amanita humboldtii (Fig. 8) is a small brown mushroom with a striate pileus margin, exannulate stipe, and a membranous, white to cinnamon colored volval sac. In the present state of knowledge, it is distinguished from *A. fuligineodisca* by microscopic characters.

We have enlarged the macroscopic information in the protologue with data from Singer's original notes (in F) on the holotype collection as well as with data from recent collections. The data on anatomy include that from a complete anatomical review of the holotype. For comparison to *A. fuligineodisca* see the discussion of that species directly above. Apparently *A. humboldtii* is a rather rare species.

The specimen *Franco-M. 192* (NY, PSO) is probably assignable to this taxon; while it is immature and lacks spores, its anatomy is identical to that of *Franco-M. 190* which was collected in the same locality.

The literature includes reference to one additional collction which we were unable to study: COLOMBIA. CUNDINAMARCA: Cerro de Las Mercedes, road between Bogotá and La Mesa, *s.d.*, *Guzmán 4602* (ENCB, COL).

6. **Amanita sororcula** Tulloss, Ovrebo & Halling, *sp. nov.* Figs. 11–12

Typus: COLOMBIA. ANTIOQUIA: Mpio. Santa Rosa de Osos, vereda "El Chaquiro," road from Santa Rosa de Osos to Aragón, near Llanos de Cuivá, 14 Nov 1986, *Ovrebo 2507* (holotypus: HUA).

= *Amanita inaurata* Secr. f. *americana* E. J. Gilb., *nom. nud.*, *sensu* Singer, *Mycopathol. Mycol. Appl.* 20: 241. 1963. [Gilbert (1940-1941) listed no type and provided no Latin diagnosis.]

Etymology: *sororcula*, little sister; epithet intended to indicate a presumedly close relation with *A. ceciliae.*

Pileus 35–60 mm latus, griseus vel fumosus brunneo-griseus vel griseo-brunneus; margo sulcatus vel pectinatus; materiae volvicae ex assumentis paucis albis vel griseis adusto-griseis formantia. Stipes 70–150 × 7–10 mm, albus, ad basim interdum cinerascens, exannulatus; volva alba vel pallida cinerea vel grisea submembranacea, infra ex assumentis ut armillis formata, interdum basi cupula minuta inclusa. Fibulae nullae. Sporae inamyloideae, globosae vel subglobosae, (8.8–)9.7–12.8(–16.8) × (8–)8.8–12(–15.5) µm.

PILEUS: 35–60 mm diam., uniformly gray to smoky gray (61 g Br) to brownish gray (7E3) to "soapstone" (MP) (grayish brown), darker gray over disk, paler toward margin, broadly convex to campanulate, subumbonate to umbonate, dry, subpolished to shiny; margin sulcate or pectinate (0.25R–0.35R); context white, but black under pileipellis, unchanging when cut, 4–5 mm thick; universal veil remnants as a few scattered, white to gray to tannish gray (5C3) appressed patches. LAMELLAE: free, white to whitish buff, with white to gray to dark gray to dark brownish gray (7E3), fimbriate edges that always turn brown on drying, drying 4A3 to slightly grayer than 4A4 to 5B4, rather thin, 6–7.5 mm broad. STIPE: 70–150 × 7–10 mm, subcylindric or tapering upwards, white (at least above), sometimes gray toward base; surface decorated with fine appressed gray to brownish gray (7E3) fibrils often arranged in belts; exannulate; context white, hollow; universal veil a complete or incomplete short limb of white to pale ashgray to gray membranous to submembranous tissue about 20–30 mm from stipe base, or in scattered gray patches or bands low on stipe, sometimes with a white, rather small portion enclosing base of stipe. Odorless. Taste not recorded. Macrochemical tests: negative for laccase; positive for tyrosinase throughout basidiocarp. Phenol (also an indicator for tyrosinase) turning flesh directly chocolate.

PILEIPELLIS: 50–55 μm thick, gelatinized just at surface; orange-brown intracellular pigment present in some hyphae, sometimes the pigmented material becoming guttulate in 2% KOH; filamentous, undifferentiated hyphae 1.8–7.5 μm diam., interwoven, branched, subradially arranged in part; oleiferous hyphae common to plentiful, 1–11.5 μm diam., interwoven. PILEUS CONTEXT: filamentous, undifferentitated hyphae 1.8–8.2 μm diam. branched; inflated cells clavate to ovoid to ellipsoid to elongate, to 86 × 40 μm; branched, oleiferous hyphae common to plentiful, locally in coils and knots, 1–9 μm diam. LAMELLA TRAMA: obscurely bilateral; central stratum distinct; filamentous, undifferentiated hyphae, 1.5–8.2 μm diam., branched; inflated cells obscure or absent below subhymenium; oleiferous hyphae scarce to locally common, 3–5.5 μm diam. SUBHYMENIUM: cellular, occasionally locally showing some branching of chains of cells or slightly inflated hyphal segments, 3 to 4 cells thick; basidia arising from inflated cells or slightly inflated short hyphal segments up to several cells deep in the subhymenium or occasionally directly from hyphae of the central stratum; inflated cells to 15.5 × 12.5 μm. BASIDIA: (46–)50–75 × 9.5–20 μm, 4- or occasionally 3- or 2-sterigmate, thin-walled, contents granular in 2% KOH; clamps not observed. UNIVERSAL VEIL: On stipe base, exterior surface: somewhat gelatinized, cells collapsed, with orange-brown intracellular pigment in some hyphal segments and some inflated cells, dominated by inflated cells. On stipe base, interior and inner surface: filamentous, undifferentiated hyphae dominant and hyaline, 1.5–8 μm diam., branched, loosely interwoven, sometimes in fascicles, with walls slightly thickened (±0.5 μm) in those of larger diam.; inflated cells thin-walled, subglobose to ovate to ellipsoid to elongate, terminal or in short chains, to 67 × 50(–62) μm; oleiferous hyphae not common, 1–2.8 μm diam. On pileus: very similar; filamentous, undifferentiated hyphae to 11.5 μm diam.; inflated cells to 72 × 52 μm; oleiferous hyphae 1.8–11 μm diam., more common than at stipe base. STIPE CONTEXT: acrophysalidic; filamentous, undifferentiated hyphae 1.5–10.5 μm diam., branched; acrophysalides to 354 × 50 μm, but most less than half that length, walls thin to 0.8 μm thick; branched oleiferous hyphae so plentiful as to partially obscure other structures at times, locally in coils and knots, 2.5–14.5 μm diam., those of largest diameter with slightly thickened walls.

BASIDIOSPORES: [160/6/4] (8.8–)9.7–12.8(–16.8) × (8–)8.8–12(–15.5) μm [**L** = 10.2–12.1 μm; **L'** = 11 μm; **W** = 9.3–11.5 μm; **W'** = 10.2 μm; Q = (1–)1.01–1.16(–1.25); **Q** = 1.05–1.10; **Q'** = 1.07], inamyloid, hyaline, thin-walled, smooth, globose to subglobose, slightly to distinctly adaxially flattened; contents guttulate; apiculus sublateral, truncate conic to cylindric, to 3.5 μm broad at the base; color in deposit not recorded.

Distribution and habitat: Solitary to subgregarious, in *Quercus humboldtii* woods in April, June and November and under *Q. colombiana* Cuatrec. and *Q. humboldtii* in July, at ±2500–3000 m elevation.

Collections examined: COLOMBIA. ANTIOQUIA: Mpio. Santa Rosa de Osos, vereda "El Chaquiro," road from Santa Rosa de Osos to Aragón, near Llanos de Cuivá, 5 Nov 1986, *Ovrebo 2426* (HUA), 22 Jun 1990, *Franco-M. 534* (HUA, NY), 30 Apr 1991, *Franco-M. 565* (HUA, NY). BOYACÁ: Tunja, Arcabuco, 29 Jul 1960, *Singer B3596* (BAFC 31.974).

Discussion

Our description is enhanced by that in Singer (1963) and by Singer's notes on *Singer B3596* in his notebooks (F).

Amanita sororcula (unillustrated) has a gray to smoky gray, unpolished pileus with a sulcate margin. The pallid to gray universal veil remnants may remain as occasional patches on the pileus or as patches or rings on the lower part of the stipe. This species belongs with those of section *Vaginatae* distinguished by a friable universal veil. For most of the past century, the invalid name "*A. inaurata* Secr." was applied to almost all such entities. Bas (1984) noted that the valid name for the European entity described by Secretan is *A. ceciliae* (B. & Br.) Bas. Now *A. ceciliae* is beginning to be used in a very broad sense to replace Secretan's name. This is because, throughout the world, few other taxa have been described with a friable universal veil and globose to subglobose spores. D. T. Jenkins, Department of Biology, University of Alabama at Birmingham (private communication) is investigating the taxa of section *Vaginatae* in North America which have friable universal veils. *Amanita sororcula* should be compared with the two species of this group having the most similar spores: *A. ceciliae* of Europe and *A. cinctipes* Corner & Bas of Singapore.

Amanita sororcula is distinguished from *A. ceciliae* by a grayer pileus, smaller habit, and its smaller, less perfectly globose spores. Notes of C. Bas filed with the type and a topotype (possibly an isotype) collection in K provide the following information for the spores of *A. ceciliae*: [20/2/2] 10.5 – 15.5(– 21) × 10.5 – 15.3(– 20) μm. The volval material described in Bas' notes in K is very similar to that described above for *A. sororcula*. Unfortunately, Bas found the hymenium of the type could not be rehydrated.

Amanita cinctipes was reported by Corner and Bas (1962) to have spores 8.2 – 11.1 × 7.8 – 10.1 μm with Q = 1.05. The pileus was reported to have more brown tints and to be very dark in the disk. The basidia of the material studied by Corner and Bas did not exceed 50 μm in length, and the subhymenium is densely ramose and without globose cells.

Amanita inaurata f. *americana sensu* Singer and possible synonyms are mentioned in Garrido and Bresinsky (1985). The synonymy proposed was with *A. gemmata sensu* Guzmán & Varela which is clearly not a member of section *Vaginatae*; in fact, from Guzmán and Varela's description (1978) there seems to be no way that their "*A. gemmata*" (which we believe to represent *A. xylinivolva*) could be confused with "*inaurata* f. *americana*," *i.e.*, *A. sororcula*. See also the discussion under *A. xylinivolva*, above. Henao-M. and Tobón-A. (1984) described four collections under the heading "*A. inaurata*;" we believe that they may be referable to *A. sororcula*.

Subgenus *Lepidella* (E. J. Gilb.) Veselý emend. Corner & Bas Section *Lepidella*

7. **Amanita advena** Tulloss, Ovrebo & Halling, *sp. nov.* Figs. 13 – 15, 33

Typus: COLOMBIA. ANTIOQUIA: Mpio. San José de la Montaña, along road from San José de la Montaña to Labores, 26 Nov 1986, *Ovrebo 2547* (holotypus: HUA; isotypus: NY).

Etymology: *advena*, a stranger, because of the apparent rareness of members of section *Lepidella* subsection *Solitariae* Bas in South America.

Pileus 85–120 mm latus, albidus vel pallide bubalinus, cinnamomescens, interdum demum areolatus; margo estriatus, appendiculatus; materiae volvicae assumenta minuta vel verrucas magnas formantes, initio albae brunnescentes. Stipes 110–130 × 15–25 mm, bubalinus, super bulbum basim ovoideum cinnamomescens;

annulus apicalis; volva balteos inconspicuos in pagina stipitis et bulbi formans. Fibulae praesentes. Sporae amyloideae, late ellipsoideae vel ellipsoideae, interdum subglobosae, (8.5–)9–12.5(–14) × (6.5–)7–9.5(–10.2) µm. In subsectione Solitariis stirpis Ravenelii cellulis numerosis inflatis subhymenii et amplitudine et forma sporarum distinguenda.

PILEUS: 85–120 mm diam., white to light buff when young, becoming tinged with cinnamon in time, not otherwise discoloring, viscid when moist, at maturity broadly convex to slightly concave, dry, becoming areolate; margin nonstriate, appendiculate, decurved at first, becoming rimose; context whitish buff, ±10 mm thick; universal veil white at first, then brown, breaking into flattened to slightly erumpent, large warts to small patches, at times in concentric rings. LAMELLAE: free, close, light cream, becoming dingy ochraceous in age near pileus margin, strongly fimbriate, 8–11 mm broad; lamellulae numerous. STIPE: 110–130 × 15–25 mm, buff, becoming cinnamon above the basal bulb,narrowing upward; surface scabrous in places, also breaking into scales; annulus at very apex, white, then browning in spots, membranous to submembranous, striate above, smooth below, margin thickened, becoming lacerate; basal bulb ovoid, tapering to a rounded point, white mycelial "threads" present on lower third; universal veil remnants white at first, browning, in a series of obscure rings from a point slightly above the top of the bulb to a point below the broadest part of the bulb. Odor slightly putrid. Taste indistinct. Macrochemical spot tests: 3% KOH light ochraceous on stipe, faintly yellow on less mature pileus.

PILEIPELLIS: yellowish brown to nearly hyaline, rather thin, 30–90 µm thick, only gelatinized at surface; filamentous, undifferentiated hyphae 1.5–10 µm diam., interwoven, branched, subradially oriented; oleiferous hyphae 2.8–9.5 µm diam., branched, locally plentiful. PILEUS CONTEXT: filamentous, undifferentiated hyphae 2.5–14.5 µm diam., branched; inflated cells subglobose, ovoid, clavate, ellipsoid, elongate, dominant, to 86 × 64 µm; oleiferous hyphae scarce, ±4 µm diam. LAMELLA TRAMA: bilateral; filamentous, undifferentiated hyphae 1.2–11.5 µm diam., some segments inflated to 18 µm diam., branched; inflated cells ellipsoid, elongate, clavate, pyriform, to 71 × 49 µm; oleiferous hyphae common, 2.2–6.2 µm diam., branched, locally plentiful in knots; clamps present. SUBHYMENIUM: dominantly cellular; basidia arising from subglobose and clavate inflated cells and from somewhat inflated hyphal elements. BASIDIA: (27.5–)41–63 × 11–17.5(–27) µm, dominantly 4-sterigmate, also 3-, 2-, and occasionally 1-sterigmate; clamps and proliferated clamps present. UNIVERSAL VEIL: On pileus: gelatinized on outer and inner surfaces; becoming orange-brown in some regions on outer surface of wart, always browner than adjacent pileipellis; alignment of elements subvertical; filamentous, undifferentiated hyphae 1.8–15 µm diam., branched, rather plentiful near upper surface, dominant near lower surface, interwoven, constricted at septa when of larger diameter; inflated cells plentiful, thin-walled, single when terminal, subglobose, subpyriform, clavate, elongate, to 122 × 53 µm; oleiferous hyphae branched, 1.8–8.5 µm diam. On stipe base and upper bulb: with same pigmentation and elements as on pileus; elements more often collapsed, but of very similar shapes and sizes; orientation of elements not discernible; inflated cells to 107 × 70 µm; oleiferous hyphae scarce. STIPE CONTEXT: acrophysalidic; filamentous, undifferentiated hyphae 1–14 µm diam., branched; acrophysalides plentiful to dominant, to 278 × 84 µm; oleiferous hyphae branched, locally in dense tangles and knots, 2.5–12.5 µm diam.; clamps present. PARTIAL VEIL: filamentous, undifferentiated hyphae 4.8–13.5 µm diam., branched, often with subradial orientation, narrow diameter elements with unconstricted septa and long segments, broader elements with constricted septa and relatively short segments; inflated cells terminal, relatively common, clavate to broadly clavate, to 94 × 47 µm; oleiferous hyphae scarce, 1.5–5.5 µm diam.

BASIDIOSPORES: [65/2/1] (8.5–)9–12.5(–14) × (6.5–)7–9.5(–10.2) µm [**L** = 10.5–10.7 µm; **L'** = 10.6 µm; **W** = 8.2–8.3 µm; **W'** = 8.2 µm; Q = (1.10–)1.11–1.57(–1.62); **Q** = 1.29–1.31; **Q'** = 1.30], amyloid, smooth, thin-walled, hyaline, more or less adaxially flattened, broadly ellipsoid to ellipsoid, occasionally subglobose, infrequently elongate, sometimes markedly swollen at one end; contents granular to guttulate; apiculus most commonly cylindric, occasionally truncate conic; color in deposit not recorded.

Distribution and habitat: In *Quercus humboldtii* forest in November, at ±2500–3000 m elevation.

Collection examined: Known only from the type.

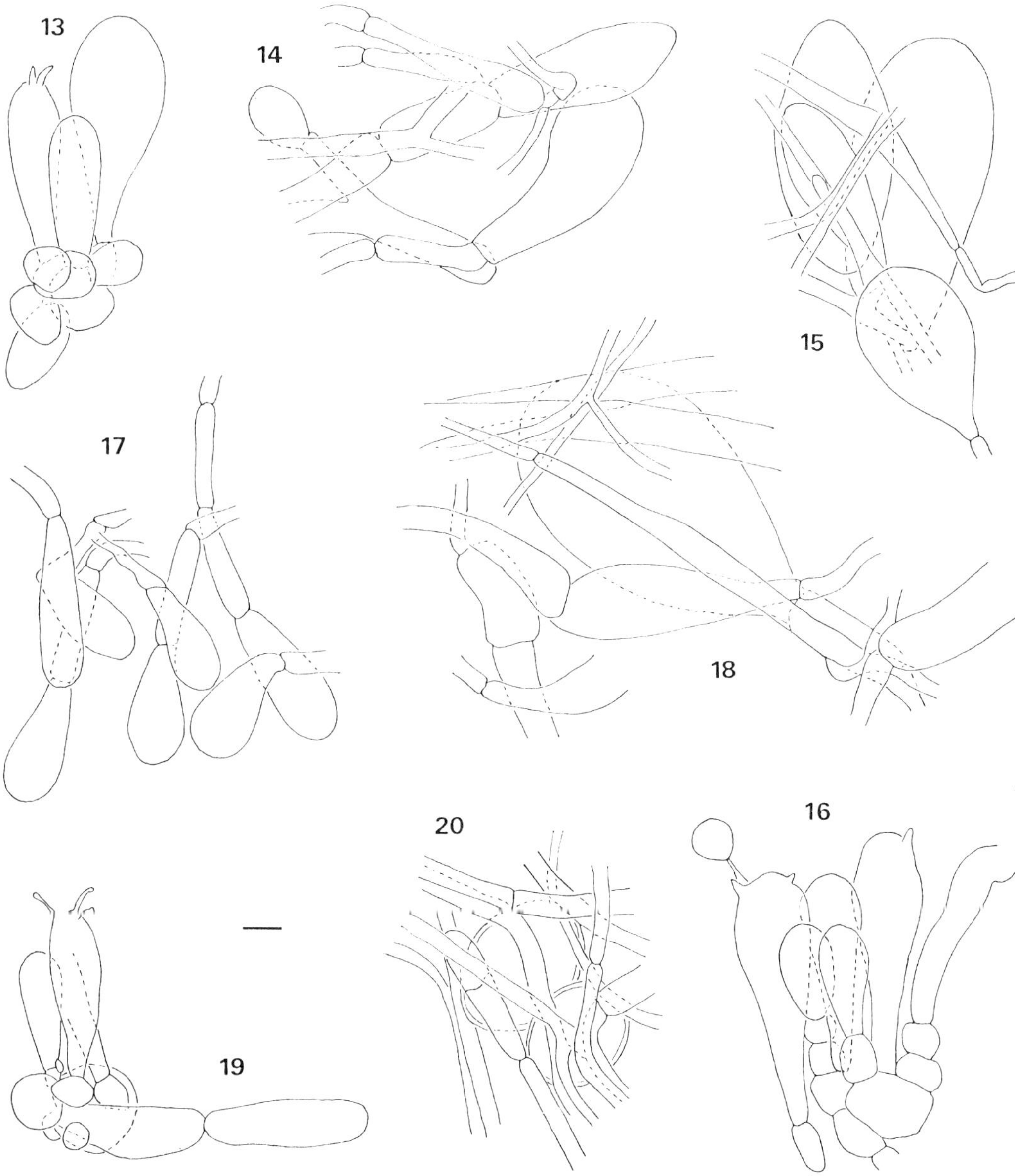

Figs. 13–20. Microscopic features. Figs. 13–15. *Amanita advena* (*Ovrebo 2547*). 13. Hymenium and subhymenium. 14. Partial veil. 15. Elements from crush mount of upper portion of universal veil from pileus wart. Figs. 16–18. *Amanita arocheae* (*Halling 5071*). 16. Hymenium and subhymenium. 17. Partial veil. 18. Universal veil from stipe base. Figs. 19–20. *Amanita aureomonile* (*Franco-M. 156*). 19. Hymenium and subhymenium. 20. Universal veil from stipe base. Scale line = 10 μm.

Discussion

Amanita advena (Fig. 33) is a medium to large mushroom which is pallid at first, then discolors to cinnamon in many parts. It has an unpleasant odor. The stipe has a bulbous base with somewhat obscure lines of universal veil material circling the top of the bulb. The basidia are occasionally clamped.

Within section *Lepidella*, *A. advena* is assignable to subsection *Solitariae*. Bas (1969) noted that members of the subsection are very rare in South America. The southeastern United States is one of the centers of distribution for the subsection. Hence, it is possible that this entity or an ancestral species entered South America with the expansion of *Quercus* from the North.

Because the elements of the volva have a pronounced subvertical alignment and filamentous hyphae dominate in the base of the warts, assignment to stirps *Ravenelii* is indicated by Bas' (1969) keys. Three species are placed by Bas in this stirps. They are discussed below. All are reported by him to have ramose subhymenia, a feature that differentiates them from *A. advena*.

—*Amanita crassa* Bas (described from Argentina) was previously the only member of subsection *Solitariae* known from South America. It may be differentiated easily by its small, narrow basidia and small spores with Q = 1.5 (Bas, 1969).

—*Amanita ravenelii* (B. & C.) Sacc. was described from eastern North America. According to Bas' (1969) observations, there are a number of characters of *A. ravenelii* which distinguish it from *A. advena*: Many or most of the elements of the universal veil of *A. ravenelii* have refractive contents; its spores are markedly narrower with Q = 1.5 – 1.6; its basidia are narrower. Tulloss has frequently collected *A. ravenelii* and has found it to have a much more rooting bulb than *A. advena* and to have an evanescent partial veil. Raithelhuber (1986) cast doubt on earlier reports of *A. ravenelii* in South America. The entity described as *A. ravenelii* by Raithelhuber (1987) is not in agreement with the North American species; *e.g.*, Raithelhuber's material has spherical spores.

—*Amanita strobilacea* (Cooke) Sacc. of Australia has globose to broadly ellipsoid spores with Q of about 1.2.

Of the species from South America described as belonging to section *Lepidella* since Bas's 1969 monograph, the only entity to which *Ovrebo 2547* might be referred is *A. morenoi* Raithelhuber (1986), described from Argentina. However, its habit was described as smaller than that of *A. advena*, and the pileus was described as pale-brownish to chestnut brown. Its stipe was reported to be cream-colored, ± cylindric, somewhat expanded at the base, and to have small, chestnut brown scales. We have examined the holotype (Argentina. Prov. Rio Negro, Parque Nacional Nahuel Huapí, 9 Mar 1959, *Gamundí 30.612*, BAFC); its spores measure (8.4 –)8.8 – 11.9(– 19.6) × 7.7 – 11.2(– 15.4) μm [Q = 1 – 1.27(– 1.31); Q = 1.14], are inamyloid, and contain a yellow amorphous substance which, in mature spores, gathers in a layer 0.8 – 1.1 μm thick on the inside of the spore wall. Raithelhuber assigned the species to section *Lepidella*; however, it belongs in section *Amanita*.

Amanita fernandeziana, *nom. prov.*, described from a single fruiting body collected in the Juan Fernandez Islands, Chile (Singer, 1959), was not assigned to a section by Singer. Since it has amyloid spores, a poorly developed pileipellis, and a volva that is absent at maturity, it could belong to section *Lepidella*. The spores are almost identical to those of *A. advena*. However, *A. fernandeziana* has all parts darkening considerably with age and is apparently much smaller than *A. advena* (Singer, 1959). We have not seen Singer's material.

Compared with species outside of South America, *Amanita advena* has spore size and shape somewhat similar to those of the recently described *A. ingrata* Pegler (from Martinique) and *A.*

pyramidifera Reid (from Australia). Based on the protologue (Pegler, 1983), the former belongs in subsection *Vittadiniae* Bas. Reid (1980) provided a number of characters that distinguish *A. pyramidifera* from *A. advena*: poorly developed annulus, narrower basidia, much smaller habit and gray coloring, only scant hyphae among subglobose cells of the universal veil, and more plentiful clamps on the basidia.

SUBGENUS *Lepidella* SECTION *Phalloideae* (Fr.) Quél.

Two species of this section are known from the region of study. It is curious that these two species are the only two members of the genus known so far to occur under *Colombobalanus* (=*Trigonobalanus*) *excelsa* (Lozano-C. *et al.*) Nixon & Crepet. Also, to our knowedge, this is only the second report of purported ectomycorrhizal associations of trigonobalanoids. In the first such report (Forman, 1963), E. J. H. Corner noted that a species of *Russula* grew around *Trigonobalanus verticillata* Forman in Borneo.

1. Pileus umber brown to brownish gray to gray, virgate; annulus white; stipe white or with pale gray tints; 75% of spores longer than 8 μm and with Q = 1–1.11(–1.27) .... 8. *Amanita arocheae*

1. Pileus and annulus bright yellow; stipe pale yellow; spores less than 8 μm long and with Q = (1–) 1.03–1.24(–1.31) .... 9. *Amanita aureomonile*

8. **Amanita arocheae** Tulloss, Ovrebo & Halling, *sp. nov.* Figs. 16–18, 34

Typus: COLOMBIA. ANTIOQUIA: Mpio. San José de la Montaña, ±13 km S of San José de la Montaña along road to Labores, 26 Nov 1986, *Halling 5071* (holotypus: HUA; isotypus: NY).

Etymology: In honor of the Mexican mycologist, R. M. Aroche, who has studied similar taxa close to *Amanita phalloides* in Mexico.

Pileus 30–110 mm latus, virgatus, initio griseus demum griseo- vel umbrino-brunneo-striatus; margo estriatus, exappendiculatus; materiae volvicae nullae. Stipes 80–200 × 5–17 mm, albus vel pallide cinerascens; annulus apicalis copiosus membranaceus, bulbo quam stipite crassiori; volva alba membranacea limbata. Fibulae nullae. Sporae dilute amyloideae vel amyloideae, globosae vel subglobosae vel interdum late ellipsoideae, (5.8–)7–10(–12.8) × (5.5–)6.8–9.5(–12) μm.

PILEUS: 30–110 mm diam., rounded subconic at first, then convex, often becoming umbonate, finally plano-convex, slightly gray, umber brown (5F6) at disc, streaked with apparent inherent radial fibrils, 5C–D4 to 5E3 to 7C2 toward margin, grayer when young, browner at maturity; margin nonstriate or very slightly so in smaller specimens, nonappendiculate; context white, 4 mm or more thick, not bruising or staining on exposure; universal veil absent. LAMELLAE: free to very narrowly adnate, occasionally with faint decurrent line or slight decurrent tooth, close to crowded, thin, white, 4 mm or more broad, with fimbriate edges; lamellulae attenuate and of varying length, frequent. STIPE: 80–200 × 5–17 mm, white with white to pale gray surface fibrils, not changing color when bruised, subcylindric or narrowing upward; context white, not bruising; bulb globose to subglobose and significantly broader than stipe; annulus white, membranous, flaring, ample, persistent, apical to subapical, eventually collapsing on stipe; universal veil limbate, white to pale gray, limb sometimes tearing and becoming appressed to lower stipe. Odor mild to disagreeable. Taste not recorded. POISONS: contains amatoxins. Macrochemical tests: negative for tyrosinase; negative for laccase; H_2SO_4 on lamellae, negative; KOH on context, negative; $FeSO_4$ on context, negative; paradimethylaminobenzaldehyde (PDAB) on context, pale vinaceous after about 30 min.; also see discussion, below.

PILEIPELLIS: filamentous, undifferentiated hyphae 2–10 μm diam., infrequently gelatinized, mostly radially-arranged, branched, many with brownish-yellow contents, gathered in fascicles, the latter not concentrated on surface; oleiferous hyphae 1.5–5 μm diam. PILEUS CONTEXT: filamentous, undifferentiated hyphae 2.4–19.2 μm diam., interwoven, branched; inflated cells terminal, elongate to ovoid, to 179 × 60 μm, thin-walled; oleiferous hyphae plentiful, 3.5–9 μm diam. LAMELLA TRAMA: bilateral, narrow; hyphae and inflated cells branching off from central stratum at a shallow angle; filamentous, undifferentiated hyphae 4.2–8.4 μm diam., branched; inflated cells ovoid, ellipsoid, cylindric, to 71 × 21 μm; oleiferous hyphae 1.5–8.8 μm diam., at times dominating the central stratum, locally in tangles, often intruding into subhymenium. SUBHYMENIUM: cellular; basidia arising directly from the inflated cells of the trama or from small inflated cells arising singly or in short chains (apparently) from inflated cells of trama. BASIDIA: 30–54 × (8.8–)10–18.2 μm, dominantly 4–, occasionally 2–sterigmate, thin-walled; clamps not observed. UNIVERSAL VEIL: Absent from pileus. At stipe base, exterior surface: longitudinally arranged fascicles of branched, filamentous, undifferentiated hyphae 3.2–14 μm diam.; oleiferous hyphae occasional, 3.5–6.3 μm diam. At stipe base, interior: dominated by densely interwoven, branched, filamentous, undifferentiated hyphae 2–10.2 μm diam., with some of those of larger diameter having walls 0.5–0.8 μm thick; inflated cells frequent, mostly thin-walled, some with slightly thickened walls, ovoid to clavate (to 105 × 50 μm) to subpyriform or subglobose (to 57 × 53 μm); branched oleiferous hyphae 4.8–7.5 μm diam. STIPE CONTEXT: acrophysalidic; filamentous, undifferentiated hyphae 3–24.5 μm diam., branched; acrophysalides narrow, to 248 × 51 μm, mostly thin-walled, occasionally with wall to 1.5 μm thick; oleiferous hyphae 1–12.6 μm diam. PARTIAL VEIL: filamentous, undifferentiated hyphae 2.5–7.2 μm diam., interwoven, branched, plentiful; small terminal cells to 42 × 15.2 μm, plentiful, clavate, subpyriform, thin-walled; oleiferous hyphae 1–21.7 μm diam., occasionally locally common.

BASIDIOSPORES: [210/10/8] (5.8–)7–10(–12.8) × (5.5–)6.8–9.5(–12) μm [L = (7.5–)8.1–9.4 μm; **L'** = 8.6 μm; **W** = (7.1–)7.6–8.9 μm; **W'** = 8.1 μm; Q = 1–1.11(–1.27); **Q** = 1.04–1.07(–1.10); **Q'** = 1.06], weakly amyloid to amyloid, hyaline, thin-walled, smooth, often adaxially flattened, globose to subglobose, occasionally broadly ellipsoid; contents guttulate to granular; apiculus sublateral, cylindrical, sometimes prominent; color in deposit not recorded.

Distribution and habitat: In Colombia, solitary, at 2500 m elevation under *Quercus humboldtii* in loamy soil in November and under *Colombobalanus excelsa* in May. In Mexico in mesophytic, montane woods (*e.g.*, *Pinus/Quercus* forests and forests including *Q. crassifolia* Humb. & Bonpl., *Q. candicans* Née, *Alnus jorullensis* H.B.K., and *Clethra mexicana* DC.) in humus in the states of Guerrero and Puebla, at 2150–2600 m elevation, July to September.

Collection examined: COLOMBIA. VALLE: Mpio. Pance, Parque Natural Los Farallones, Centro de Educación Ambiental El Topacio, 30 May 1990, *Franco-M. 507* (CUVC, NY).

Extralimital collections examined: MEXICO. GUERRERO: Mpio. Chilpancingo de los Bravo, Omiltemi, El Faisal, 11 Jul 1985, *Cifuentes Blanco 2138* (FCME as *A. phalloides*); Mpio. Chichihualco, km 4.5 between El Carrizal and Atoyac, 16 Aug 1980, *Madariaga s.n.* (FCME 1152 as *A. phalloides*). PUEBLA: Mpio. Teziutlán, cañada de Jalapasquito, desviación a Ixtlahuaca, [Sep? 1983?], *Aroche 517* (XAL as *A. phalloides*); Mpio. Tezuitlán, cerro Techachaca, 27 Aug 1983, *Aroche & Xolalpa 315* (FCME as *A. phalloides*), 28 Aug 1983, *Aroche & Xolalpa 317* (FCME as *A. phalloides*), 17 Sep 1983, *Aroche & Xolalpa 513* (FCME as *A. phalloides*), [Sep? 1983?], *Aroche & Bonavides 515* (FCME as *A. phalloides*).

Discussion

Amanita arocheae (Fig. 34) is very similar in habit to *A. phalloides*, but has a gray to umber brown, virgate pileus. There are two forms of *A. phalloides* reported from South America: f. *phalloides* (possibly imported from Europe) and f. *quercus-ilicis* Herter. Both reportedly have green tones on the pileus (Herter, 1934). The spores of *A. arocheae* are broader, and as a consequence, rounder than those of *A. phalloides* which have widths in the range 6.8 – 7.8 μm and Q' = 1.21 according to Jenkins (1986). In addition, the universal veil of *A. arocheae* largely lacks the thick-walled elements that are frequent in the universal veil of *A. phalloides* as noted by Reid (1980) and Tulloss and Wright (1988). There are two pallid, subspecific taxa of *A. phalloides* described from France (f. *ochroleuca* Forquignon and var. *larroquei* Massart & Beauvais, *nom. inval.*, holotype not specified). Although the anatomy of these taxa is unknown, they differ from *A. arocheae* in pileus color, spore shape, and habit. Forma *ochroleuca* has a narrow ellipsoid bulb; and var. *larroquei* has an elongated stipe, the length of which is 3 – 4 times the cap diameter (Massart & Beauvais, 1975; Massart, 1984).

Among other species of section *Phalloideae* with brown pileus and spore size and shape similar to that of *A. arocheae*, the most similar are *A. alauda* Corner & Bas and *A. privigna* Corner & Bas, both described from Singapore (Corner & Bas, 1962). Both differ from *A. arocheae* by being smaller and not having as distinctly bulbous a stipe base. In addition, *A. privigna* has a gray annulus and slightly smaller spores, while *A. alauda* has a pure white stipe and white volva without gray tones, spores with a greater tendency to be globose, inflated cells of the universal veil tending to be globose, and an unusual pileipellis marked by dark brown hyphae embedded in a gelatinous matrix in its upper layer (Corner & Bas, 1962).

Aroche *et al.* (1984) reported on what they interpreted as a new variety of *A. phalloides* which they collected in the Mexican states of Puebla, Hidalgo, and Veracruz. This entity was called *hongo gris* by inhabitants of the area around Teziutlán who did not collect it for food because they held it to be poisonous. Aroche *et al.* carried out chromatographic studies which showed that collections from all three states contained α- and β- amatoxins; in addition, a collection from the state of Puebla contained phalloidin and other amatoxins matching those extracted from European *A. phalloides*. They did not name *hongo gris* formally, but provided a macroscopic description and a description of the spores. Among material evaluated by Aroche *et al.* (1984) were three specimens we have examined: *Aroche & Xolalpa 315*, *317* and *Aroche & Bonavides 515*. The Mexican material we examined is anatomically identical to *Halling 5071*.

The very wide range of pileus colors and the difference in chromatographic results between the Puebla specimen and the other Mexican specimens suggests Aroche *et al.* (1984) may have been dealing with two taxa. It is for this reason that we reserve an enumeration on the macrochemical reactions reported by Aroche *et al.* to this point in the discussion. These reactions were: Meixner test (Vergeer, 1986) positive for amatoxins; 3% and 10% KOH negative on basidiocarp surfaces; 10% KOH pale yellow on the context; H_2SO_4 on the lamellae, lilac to purple; phenol-aniline, purple-brown on the stipe surface; HNO_3, KCl, $FeSO_4$ all negative. The positive spot test with phenol-aniline possibly contradicts the negative indication for tyrosinase in *Franco-M. 507*. However, Tulloss has observed a positive tyrosinase reaction in remnants of universal veil (limbus internus) on the stipe surface in some taxa for which the context of the stipe in a mature basidiocarp is commonly not reactive.

Franco-M. 507 is immature; however, the characteristic structures of the subhymenium, lamella trama, and universal veil were all noted in it.

Additional Mexican material is located in ENCB and SEDUE according to Aroche *et al.* (1984).

9. **Amanita aureomonile** Tulloss & Franco-Molano, *sp. nov.* Figs. 19 – 20

Typus: COLOMBIA. VALLE: Mpio. Pance, Parque Natural Los Farallones, above El Topacio, 17 Nov 1988, *Franco-M. 156* (holotypus: CUVC).

Etymology: *aureus*, golden + *monile*, necklace; from the small, yellow partial veil.

Pileus 20–25 mm latus, laete flavus; margo leviter striatus; materiae volvicae nullae. Stipes usque ad 45 × 3 mm, pallide flavus; annulus apicalis, laete flavus, membranaceus; volva alba membranacea, in bulbo minuto limbata. Fibulae praesentes. Sporae amyloideae, globosae vel subglobosae vel late ellipsoideae, (5–) 5.5–7.5(–7.8) × (4.8–)5–6.5(–6.8) μm.

PILEUS: 20–25 mm diam., bright yellow (3A6), darker at disc, moist, shiny; margin somewhat striate, decurved; context concolorous with pileipellis, thin (less than 1 mm thick); universal veil absent. LAMELLAE: free, close, white, drying slightly sordid pale yellow, narrow, ±2 mm broad, occasionally forking; lamellulae present. STIPE: ±45 × ±3 mm, pale yellow (3A2), shiny, smooth, subcylindric; annulus bright yellow (3A8), membranous, skirtlike, superior; bulb subglobose; universal veil forming a white, fleshy, limbate volva.

PILEIPELLIS: filamentous, undifferentiated hyphae 1.5–5 μm diam., radially arranged, branched, some intercalary segments slighty swollen/ventricose; no oleiferous hyphae observed. PILEUS CONTEXT: filamentous, undifferentiated hyphae 2–5.5 μm diam., loosely interwoven, branched; inflated cells elongate, thin-walled or with walls thickened slightly (to 0.5 μm thick), to 84 × 36 μm; oleiferous hyphae not common, to 12 μm diam. LAMELLA TRAMA: bilateral; filamentous, undifferentiated hyphae 1.2–5.8 μm diam., mostly restricted to central stratum; inflated cells with a very low angle of divergence, many simply intercalary to hyphae of central stratum, sometimes in chains, to 83 × 21 μm; oleiferous hyphae ±6 μm diam., not common. SUBHYMENIUM: cellular; composed of subglobose to subellipsoid to irregular cells, to 21 × 20 μm; basidia arising from these or from very short uninflated hyphal segments developing from the inflated cells. BASIDIA: 31–41 × 7.5–10.8 μm, thin-walled, 4-sterigmate; clamps relatively common. UNIVERSAL VEIL: On stipe base, exterior surface: gelatinized extensively; filamentous, undifferentiated hyphae 2–4.8 μm diam., interwoven, branched; oleiferous hyphae ±2.2 μm diam., uncommon. On stipe base, interior: unfilamentous, differentiated hyphae 1.8–8.8 μm diam., branched, dominant, many sublongitudinally oriented; inflated cells in scattered groups, with walls 0.5–0.8 μm thick, globose to ovoid to elongate (to 64 × 53 μm, usually smaller) and clavate (*e.g.*, 36 × 18 μm); oleiferous hyphae uncommon, ±3.8 μm diam. On stipe base inner surface: gelatinized extensively; apparently composed of longitudinally oriented, filamentous, undifferentiated hyphae. On pileus: absent. STIPE CONTEXT: acrophysalidic; filamentous, undifferentiated hyphae 2.8–11.5 μm diam., branched; acrophysalides thin-walled, to 175 × 47 μm; oleiferous hyphae 2–6.8 μm diam. PARTIAL VEIL: upper surface extensively gelatinized with scattered subglobose to ovoid inflated cells, *e.g.*, 42 × 32 μm; interior dominated by filamentous, undifferentiated hyphae 1–8 μm diam., subradially arranged, branched; inflated cells very scarce, clavate, 37 × 23 μm; oleiferous hyphae, scattered, 1.2–5 μm diam.

BASIDIOSPORES: [60/2/1] (5–)5.5–7.5(–7.8) × (4.8–)5–6.5(–6.8) μm [**L** = 6.6–6.7 μm; **L'** = 6.7 μm; **W** = 5.8–6 μm; **W'** = 5.9 μm; Q = (1–)1.03–1.24(–1.31); **Q** = 1.12–1.15; **Q'** = 1.13], amyloid, hyaline, thin-walled, smooth, slightly adaxially flattened, globose to subglobose to broadly ellipsoid; contents guttulate; apiculus sublateral, small, truncate conic to cylindric; color in deposit not recorded.

Distribution and habitat: In November, subgregarious, in *Colombobalanus excelsa* forest, at 1900–2000 m elevation.

Collection examined: Known only from the type.

DISCUSSION

This small, yellow member of section *Phalloideae* with a deep yellow annulus, pale yellow stipe, white, limbate, universal veil on a subglobose bulb is distinct from all entities described in that section to date. Based on the unusually small, subglobose spores (**L'** = 6.7 μm, **W'** = 5.9 μm), the taxa with greatest similarity are *A. elephas* Corner & Bas (Singapore), *A. subjunquillea* Imai (Japan), and "species 3" of Corner and Bas (1962) (Singapore).

—*Amanita elephas* - Corner and Bas (1962) gave a very full description and illustration of this mushroom. It has a dark to pale sepia or sepia-gray pileus, a white stipe, and a white to cream annulus; it is nearly three times the size of *A. aureomonile*, has a pileipellis 400–800 μm thick, and has unclamped basidia.

—*Amanita subjunquillea* - This is the most similar, previously described species. It is superbly illustrated in a watercolor by Imai (Gilbert, 1940-1941: tab. 42) and more recently in Imazeki and Hongo (1987: pl. 30, no. 212) and Imazeki *et al.* (1988: 156). Macroscopically, *A. subjunquillea* differs from *A. aureomonile* by having brown or umber tones in the yellow pileus, and an apical to subapical partial veil that is white on the upper surface, and by having a stipe to 110 mm long with brown fibrillose scales on it (Imai, 1938). In the protologue, Imai (1933) described the spores as 7.5–8 × 7 μm; Gilbert (1940-1941: pl. xxxii, fig. 2) provided scale drawings of the spores from which dimensions of 6.5–10.5 × 5.5–9 μm can be determined.

—Corner & Bas "species 3" - Corner and Bas (1962) described this entity briefly: the hyphae of the pileipellis contain a gray or umber sap; the pileus is leaden gray; the stipe is pale grayish white; basidiocarps are much larger than the specimens seen of *A. aureomonile* (nearly four times the size).

There are three species described by Patouillard from Madagascar (*A. thejoleuca, A. alliiodora,* and *A. murinacea*) and one by Hariot and Patouillard from Vietnam (*A. manginiana*) in which the spores just exceed the range found for *A. aureomonile*; however these all have gray or brown pilei and no hint of yellow on stipes or partial veils (Gilbert, 1940-1941). Moreover, Gilbert (1940-1941) generally found the spores of the types to be larger than reported by the original authors.

SUBGENUS *Lepidella* SECTION *Validae* (Fr.) Quél.

1. Pileus black to blackish-brown; universal veil gray, rather thick; spores with Q = (1.11–)1.28–1.6 (–1.74) .... 13. *Amanita picea*

1. Pileus not black to blackish-brown; universal veil pallid, cream, or yellow.

    2. Basidiocarp bruising deep red, reddish brown or wine-colored in all parts; universal veil not yellow; spores with Q = (1.05–)1.15–1.46(–1.69) .... 10. *Amanita brunneolocularis*

    2. Basidiocarp not changing color when bruised; pileus orange-brown or mustard yellow; stipe yellow to orange-brown; universal veil yellow, often lacking on pileus.

        3. Pileus brownish orange; spores (5.2–)6.8–9(–11.5) × (4.5–)5.8–7.8(–9.5) μm; Q = (1–)1.05–1.38 (–1.73) .... 11. *Amanita flavoconia* var. *inquinata*

        3. Pileus mustard or dingy yellow; spores (6–)6.2–8(–9) × (4.8–)5.8–7(–8.2) μm; Q = (1–)1.03–1.25(–1.37) .... 12. *Amanita flavoconia* var. *sinapicolor*

10. **Amanita brunneolocularis** Tulloss, Ovrebo & Halling, *sp. nov.* Figs. 21–23, 35

Typus: COLOMBIA. ANTIOQUIA: Mpio. Santa Rosa de Osos, vereda "El Chaquiro," road from Santa Rosa de Osos to Labores, near Llanos de Cuivá, 14 Nov 1986, *Ovrebo 2506* (holotypus: HUA).

Etymology: *brunneolocularis*, brown-celled, a reference to the inflated cells of the universal veil which contain a brown pigment.

Pileus 45–130 mm latus, obscure adusto-bubalinus vel incarnato-griseus vel ±rubro-brunneus, contusus rubescens; margo estriatus, annuli rudimentis interdum appendiculatis; materiae volvicae assumenta vel squamulas formantes, albidae vel griseae vel brunneo-fuscae vel rubro-brunneae, cellulis inflatis pigmentum brunneum intracellulosum saepe continentibus. Stipes 65–150 × 8–30 mm, albidus, fibrillis brunneis tenuibus, contusus rubescens, basi subbulbosus vel bulbosus; annulus superus, submembranaceus vel membranaceus; materiae volvicae nullae aut e super bulbum crista partiali constantes. Fibulae nullae. Sporae amyloideae, late ellipsoideae vel ellipsoideae, interdum subglobosae, (6.5–)7.5–9.8(–10.8) × (5–)5.8–7.8(–9) µm.

PILEUS: 45 – 130 mm diam., dull drab tan-buff to flesh-gray (near 7C3) to somewhat red brown (6E5), discoloring red where bruised, becoming broadly convex to plane with low, broad umbo, dry, subpolished, subvirgate (with 10× lens), at times areolate on disk; margin nonstriate, uplifted at maturity, appendiculate with large pieces of partial veil; context 3.5 – 8 mm thick, white to buff in disk with pink or red tinge; universal veil as pointed squamules to 1 mm high over disk, elsewhere in patches, white to gray to brownish fuscous or reddish brown (6E5). LAMELLAE: free but not remote, crowded, pure white to whitish buff, slowly discoloring reddish brown where damaged, developing brown (7D6 to 7E7) spots in age, drying to 10YR 7/6, rather thick, 3.5 – 10 mm broad, edge fimbriate and becoming reddish brown (with 10× lens); margins sometimes covered in part by large patches of partial veil; lamellulae not in regular tiers. STIPE: 65 – 150 × 8 – 30 mm, cylindric, with a fine coating of brown (6B3 to 7C4 to 8D5) appressed fibrils over white ground, at base with reddish gray (10B3) fibrils on pallid ground, sometimes splitting transversely producing rings of recurving scales, discoloring dull red to faintly pink to vinaceous in areas of damage; base subbulbous to bulbous; partial veil superior, membranous to submembranous, white to buff on upper surface, white to sordid below, easily torn from stipe and then covering young lamellae in large patches; context becoming partially hollow, white to buff, at center discoloring red in places when cut, worm holes reddish brown; universal veil absent to almost so except for a partial, rounded ridge above bulb often becoming the rim of a submarginate narrowly rounded conic bulb in exsiccati. Odor mild. Taste nondescript. Macrochemical spot tests: KOH negative. Positive for tyrosinase in pileus and stipe contexts; negative for laccase.

PILEIPELLIS: 50 – 60 µm thick, pale brown throughout, palest near interface to pileus context; filamentous, undifferentiated hyphae 1.5 – 7.5 µm diam., branched, interwoven, subradially arranged, extensively gelatinized at surface; oleiferous hyphae present, 1 – 3.2 µm diam., not uncommon locally. PILEUS CONTEXT: filamentous, undifferentiated hyphae 1.8 – 13(– 17) µm diam., branched, interwoven, with some intercalary segments slightly inflated, with some of larger diameter with slightly thickened walls; inflated cells plentiful, pyriform, ovoid-ellipsoid, elongate, clavate, to 104 × 58 µm; oleiferous hyphae uncommon, 3 – 4.2 µm diam. LAMELLA TRAMA: bilateral; angle of divergence of inflated cells mostly under 30°; central stratum distinct, but not broad; filamentous, undifferentiated hyphae 1 – 8 µm diam., branched; inflated cells broadly ellipsoid to ellipsoid to elongate, to 45 × 23 µm; oleiferous hyphae not observed. SUBHYMENIUM: cellular, 3 – 4 cells deep; basidia arising from these cells or occasionally from short uninflated hyphal segments arising from inflated to slightly inflated cells, sometimes branched (subcellular); inflated cells subglobose to ellipsoid, to 18 × 13 µm, with major diam. at times nearly perpendicular to central stratum. BASIDIA: (20 –)23 – 40 × (8.2 –)8.8 – 12.5 µm, dominantly 4-, but occasionally 2- or 1-sterigmate, thin-walled; sterigmata often short and slender, occasionally 3.8 µm long, or to 10 µm long; no clamps observed. UNIVERSAL VEIL: On pileus: gelatinized tissues becoming orange-brown; filamentous, undifferentiated hyphae 1.8 – 7.5 µm diam., frequently branched, with walls often 0.5 µm thick except in the narrowest ones; inflated cells, terminal, single or in pairs, globose to pyriform (to 78 × 66 µm with walls 0.8 – 1.2 µm thick) or ovate to ellipsoid (to 64 × 52 µm, thin-walled) or elongate-ventricose (to 64 × 25 µm, thin-walled), some with brown intracellular pigment; oleiferous hyphae scarce or absent, to ± 1.5 µm diam. On stipe base: very similar to remains on pileus, but hyphae relatively more plentiful; filamentous, undifferentiated hyphae generally narrower than on pileus; oleiferous hyphae 2.5 – 10.8 µm diam., more frequent than those on pileus. STIPE CONTEXT: acrophysalidic; filamentous, undifferentiated hyphae 1.5 – 12 µm diam., branched, sometimes ochraceous brown near surface; acrophysalides to 265 × 51 µm, most half that width, slender, with walls 0.5 – 1 µm thick; oleiferous hyphae not observed. PARTIAL VEIL: filamentous, undifferentiated hyphae 1.5 – 8.1 µm diam., branched, without any dominant orientation, sometimes coiling and tangling in knots with relatively short segments, sometimes in fascicles, sometimes with infrequent septa and rarely branched (these distinctly radially arranged); inflated cells scattered, broadly ellipsoid to clavate to broadly clavate to pyriform, terminal, to 38 × 25 μm, thin-walled or with walls to 0.5 μm thick; oleiferous hyphae not observed.

BASIDIOSPORES: [180/7/4] (6.5–)7.5–9.8(–10.8) × (5–)5.8–7.8(–9) μm [**L** = 8.5–9 μm; L' = 8.7 μm; **W** = 6.3–7 μm; **W'** = 6.7 μm; Q = (1.05–)1.15–1.46(–1.69); **Q** = 1.25–1.36; **Q'** = 1.31], amyloid, hyaline, thin-walled, smooth, adaxially flattened to slightly adaxially flattened, mostly broadly ellipsoid to ellipsoid, occasionally subglobose, occasionally slightly expanded at one end or pip-shaped; contents guttulate; apiculus sublateral, cylindric to truncate subconic; color in deposit not recorded.

Distribution and habitat: In Colombia, gregarious under *Quercus humboldtii* and in mixed forest with *Q. humboldtii*, May through June and November, to ±2500 m elevation. In Costa Rica in June, gregarious in mixed woods with *Quercus* sp. present.

Collections examined: COLOMBIA. ANTIOQUIA: Mpio. Santa Rosa de Osos, vereda "El Chaquiro," road from Santa Rosa de Osos to Aragón, near Llanos de Cuivá, 12 Jun 1990, *Franco-M. 524* (HUA, NY). BOYACÁ: Mpio. San Miguel de Sema, road from Simijacta to San Miguel de Sema, 9 May 1987, *Halling 5247* (COL, NY).

Extralimital collection examined: COSTA RICA. CARTAGO: Along Hwy. 2, Cerro de la Muerte, 22 Jun 1986, *Ovrebo 2002* (CSU).

DISCUSSION

Amanita brunneolocularis (Fig. 35) is a medium-sized mushroom with a drab tan to flesh-gray pileus and pallid to brownish fuscous to reddish brown volval warts. The context of *A. brunneolocularis* bruises reddish brown or vinaceous throughout. In these characters and in habit it is similar to *A. rubescens* of Europe. However, there are distinctive microscopic differences: the universal veil contains inflated cells many of which have a brown intracellular pigment; the inflated cells of the lamella trama are smaller than those in the European species; the spores have **Q** = 1.27–1.36 in the Colombian species as compared to 1.38–1.58 in *A. rubescens*.

Jenkins (1986) provided 6.8–9.2 × 5.2–6.8 μm for the size of spores of North American collections assigned to *A. rubescens*; and he stated **Q'** = 1.34. Spores of similar size are reported for this taxon in North America (Kauffman, 1918), in England (Phillips, 1981), in northern Europe (Lange & Hora, 1963), in central Europe (Moser, 1983), in Argentina (Raithelhuber, 1986) and in Australia (Reid, 1980). Smith *et al.* (1979) described narrow spores in North American material as 5–6 μm wide.

Kühner and Romagnesi (1974) presented a spore size range (for material from Europe and North Africa) more in conformance with what is found in *A. brunneolocularis*: 7.7–11 × 5.7–7.5 μm. Bertault (1965) and Marchand (1971), whose collecting areas are largely restricted to southwestern Europe and northern Africa, gave similar ranges: (in composite) 8–10 × 6–7.5 μm. In Chile, *A. rubescens* is reported with spores (8–)9–11 × 6–7 μm (Garrido & Bresinsky, 1985).

Pomerleau (1980) gave a range that subsumes almost entirely those of the other authors cited: 7–11 × 4–7 μm. It is not clear whether this range was developed entirely from observation on material from Québec or from specimens gathered from a broad geographic area.

Tulloss had the opportunity to collect and examine European material of *A. rubescens* from under *Fagus sylvatica* L. in the Province of Zuid Holland, Netherlands (*9-19-87-D*, RET) and from the

Grampian Region, Scotland (*9-6-88-F*, RET). He also reviewed material of that species he collected under *Pinus* in Scotland (*9-6-88-B*, RET). Basidiospore characters from these collections are given below.

BASIDIOSPORES of northern European collections of *A. rubescens* from under *F. sylvatica*: [70/2/2] (8.2–)8.8–11(–12) × (5.2–)5.5–6.8(–7.8) μm [**L** = 9.1–10.1 μm; **L'** = 9.7 μm; **W** = 6–6.4 μm; **W'** = 6.2 μm; Q = (1.28–)1.40–1.69(–1.83); **Q** = 1.53–1.58; **Q'** = 1.55], amyloid, hyaline, thin-walled, smooth, ellipsoid to elongate, infrequently broadly ellipsoid, adaxially flattened, sometimes adaxially indented; contents guttulate; apiculus small, sublateral, cylindrical; white in deposit; color in deposit not recorded.

BASIDIOSPORES of Scottish collections of *A. rubescens* from under *Pinus*: [60/2/1] (7–)8–10.2(–11.2) × (5.2–)5.5–7(–7.2) μm [**L** = 8.6–9.5 μm; **L'** = 9 μm; **W** = 6.1–6.3 μm; **W'** = 6.2 μm; Q = (1.21–)1.29–1.76(–1.87); **Q** = 1.38–1.56; **Q'** = 1.47], amyloid, hyaline, thin-walled, smooth, broadly ellipsoid to ellipsoid to elongate, often adaxially flattened; contents guttulate to granular; apiculus sublateral, cylindrical; white in deposit.

The value of **Q** is higher for spores of the European material than in *A. brunneolocularis*. The universal veil of the material from Europe differs from that of the Colombian collections by completely lacking inflated cells with brown intracellular pigment (having volval warts colored instead by orange-brown, collapsed and gelatinized cells and clumps of similarly deteriorated hyphae); by having somewhat larger (to 135 × 94 μm), subglobose to ellipsoid to elongate to clavate cells which are much more frequently in chains of two or three; by having relatively common oleiferous hyphae, 1.8–9.1 μm diam.; and by having broader, branched, filamentous, undifferentiated hyphae 1.8–13.5 μm diam. The European material differs also by having inflated cells in the lamella trama that are nearly twice the size (to 93 × 38 μm) of those in the Colombian entity and having a slightly thicker subhymenium (3–5 cells deep). The closeness of the relationship between the two entities is emphasized by the lack of other significant differences in the hymenium, subhymenium, and lamella trama.

In the field, all collections of *A. brunneolocularis* examined were identified as *A. rubescens*. In South America, *A. rubescens* has been reported under *Pinus* from Argentina, Chile, and Colombia (Raithelhuber, 1986). *Pinus* is not native to South America, so we suspect that *A. rubescens* has been imported on the roots of nursery tree stock. Investigators collecting in such environments in Colombia should be prepared for such a possibility. For the moment we leave open the question of the proper taxonomic treatment of the various entities to which the name *A. rubescens* is applied in North America.

The collection *Franco-M. 537* (HUA) consists of a single immature basidiocarp with partial veil still in place over the lamellae. The collector's notes and a microscopic investigation suggest that it belongs in *A. brunneolocularis*.

Henao-M. and Tobón-A. (1984) described one specimen under *A. rubescens*, but we have not seen their material. However, it is possible that it might be *A. brunneolocularis*.

11. **Amanita flavoconia** Atk. var. **inquinata** Tulloss, Ovrebo & Halling, *var. nov.* Figs. 24–26, 32

Typus: COLOMBIA. ANTIOQUIA: Mpio. San José de la Montaña, ±13 km S of San José de la Montaña along road to Labores, 26 Nov 1986, *Halling 5067* (holotypus: HUA; isotypus: NY).

Etymology: *inquinare*, to tarnish, because of the pileus color being more brown than in the type of the species.

A typo pileo magis brunneo et sporis latioribus differt.

PILEUS: 25–60 mm diam., orange-brown to fulvous (6–7E8) when young, becoming paler: orange-yellow to yellow-orange to tan to orangish brown (4A5–6 to 4B7 to 5A–B6 to 5A8 to darker than 5B8 to 6C8 to 6D4), darkest at disk with color at times appearing to be in inherent radial fibrils, not discoloring, hemispheric to rounded conic to umbonate, eventually convex to plane, subviscid when moist; context 2–4 mm thick, white, yellowish orange (including 3A4) to concolorous with the pileus near pileipellis; margin nonstriate at first, sometimes becoming slightly striate in age, nonappendiculate; universal veil remnants absent or in small, detersile, submembranous, minutely verruculose, yellow (4A5 to 4A–B8 to 5B4) patches becoming pale and brown slightly with age and desiccation. LAMELLAE: free to rather distant from stipe, close, white with orange yellow to yellow (3A4) edge at least when young, with edge sometimes turning white at maturity, drying to 7.5YR 8/6 to 7.5YR 7/8, 3.5–6 mm broad; lamellulae present, of varying lengths, not in regular tiers, truncate. STIPE: 60–150 × 4–15 mm, white or pallid in part, but usually with yellow (3–4A5–6) to orange regions above and/or below the annulus, narrowed slightly upward, flared at apex, shiny to silky, with fine longitudinal striations; context white, solid at first, becoming hollow; bulb stark white, subglobose, slightly flattened longitudinally, sometimes faintly marginate or subabrupt; annulus superior to submedian, thin, membranous, persistent, collapsing on stipe, yellow (4A8 to 5B8) to yellowish tan or yellowish brown below, becoming brown with age, striate above, smooth below; universal veil in yellow (3A3) patches like those on the pileus, easily lost in collecting, or as a thin yellow rim around top of bulb. Odor and taste absent. Macrochemical tests: negative for laccase; negative or positive only in lamellae and in context of pileus for tyrosinase.

PILEIPELLIS: filamentous, undifferentiated hyphae 2.4–5.9 μm diam., subradially arranged, branched, gelatinized, more deeply yellow in lower regions where diameter reaches 22.4 μm; oleiferous hyphae infrequent, 1.8–4.9 μm diam. PILEUS CONTEXT: filamentous, undifferentiated hyphae 2–18 μm diam., branched, some segments slightly inflated; inflated cells to 89 × 51 μm; oleiferous hyphae 1–3.2 μm diam., uncommon. LAMELLA TRAMA: bilateral, with shallow angle of divergence; filamentous, undifferentiated hyphae 3.2–17 μm diam., branched, occasional segments slightly inflated; inflated cells allantoid to clavate, ovoid, ellipsoid, subglobose, subpyriform, to 109 × 63 μm; oleiferous hyphae 1.8–10 μm diam., plentiful, moderately to sparsely branched. SUBHYMENIUM: cellular; basidia arising from very small inflated cells and these, in turn, arising in small groups or clusters from the surface of larger inflated cells. BASIDIA: 16.5–40 × 6.5–13.2 μm, dominantly 4-sterigmate, but occasionally 2- or 3-sterigmate, thin-walled; clamps not observed. UNIVERSAL VEIL: On pileus: branched, filamentous, undifferentiated hyphae sparse, 2.3–11 μm diam.; dominated by inflated cells, terminal, single or (infrequently) in pairs, to 80 × 69 μm, thin-walled, occasionally deeper yellow than other elements; oleiferous hyphae 2–7.2 μm diam., locally clustered, some hyphal segments more deeply yellow than most. Absent from stipe base. STIPE CONTEXT. acrophysalidic; filamentous, undifferentiated hyphae 2.5–9 μm diam., branched; acrophysalides dominating, to 285 (or longer) × 49 μm; oleiferous hyphae relatively scarce, 3.2–4.8 μm diam. PARTIAL VEIL: almost entirely composed of branched, interwoven, filamentous, undifferentiated hyphae 2–15 μm diam., with a majority radially arranged; inflated cells scarce, terminal or in short terminal branched chains, to 63 × 23 μm; oleiferous hyphae, 1.5–3.5 μm diam.

BASIDIOSPORES: [485/23/6] (5.2–)6.8–9(–11.5) × (4.5–)5.8–7.8(–9.5) μm [**L** = (7.3–)7.5–8.4(–8.8) μm; **L'** = 7.9 μm; **W** = 6–7(–7.1) μm; **W'** = 6.6 μm; Q = (1–)1.05–1.38(–1.73); **Q** = 1.12–1.28; **Q'** = 1.20], amyloid, hyaline, thin-walled, smooth, globose to subglobose to broadly ellipsoid; contents guttulate; apiculus sublateral, cylindric, occasionally rather broad; color in deposit not recorded.

Distribution and habitat: Gregarious under *Quercus humboldtii* at 1900–2500 m (or higher) elevation, in loamy soil, April to June and November.

Collections examined: COLOMBIA. ANTIOQUIA: Mpio. Guarne, near Centro Experimental Piedras Blancas, ± 14 km E of Medellín, 10 Nov 1988, *Halling 6078* (HUA, NY); Mpio. Santa Rosa de Osos, vereda "El Chaquiro," road from Santa Rosa de Osos to Aragón, near Llanos de Cuivá, 25 Nov 1986, *Ovrebo 2537* (CSU, HUA, NY), 13 May 1987, *Halling, Mueller & Strack* [*GMM 2819*] (F 1080754, mixed collection), 12 Jun 1990, *Franco-M. 519* (HUA, NY). CUNDINAMARCA: Between Pacho and Páramo Guerrero, 9 Apr 1968, *Singer B6029* (F 1014373). NARIÑO: Mpio. La Florida, ±27 km from Pasto, Hacienda El Barranco (prop. F. Villareal), 11 May 1990, *Franco-M. 445* (NY, PSO).

DISCUSSION

Amanita flavoconia var. *inquinata* (Fig. 32) is a small to medium-sized entity with orange-brown pileus; white to yellow to orange stipe having a small bulb at the base; yellow, membranous partial veil; and yellow, detersile universal veil material. *Amanita flavoconia* var. *inquinata* differs from *A. flavoconia* var. *flavoconia* by having brown tones in the pileus coloration; moreover, var. *inquinata* has Q' = 1.20 as opposed to 1.44 for var. *flavoconia* (Jenkins, 1986). In all other characters we examined, the two appear extremely similar if not identical. In a genus in which interfertility studies have been impossible due to the extreme difficulty of growing the species in culture, the decision to describe a new taxon at the species or a subspecific level is guided by precedent and, if possible, consensus of experts. Bas (1969) and Jenkins (1977) both chose the level of variety in cases when there was some difference in spore shape or size combined with one distinguishing character such as pileus coloration or a marked bruising or staining reaction. For example, see the varieties of *A. muscaria* in Jenkins (1977) and the varieties of *A. cinereoconia* Atk. or *A. rhoadsii* (Murr.) Murr. in Bas (1969). We follow these precedents.

There are three less well-known, North American taxa, apparently closely related to *A. flavoconia*, with yellow to orange pileus and similarity of microscopic structures: The holotype of *A. elongata* Peck (from Pennsylvania) was studied by Jenkins (1978); macroscopically, this species is very close to pale striped collections of *A. flavoconia* var. *flavoconia*, but the spores have Q' = 1.51. *Amanita flavivolva* Murr. and *A. fraterna* (Murr.) Murr. (known from Florida) both have a gray disk on a pileus otherwise a shade of yellow, a white annulus, and spores with Q' greater than 1.45 (Jenkins, 1986). Spore shape and cap pigmentation distinctly separate *A. flavoconia* var. *inquinata* from all these taxa.

The northern limit of distribution of this taxon may be in Mexico or the southern United States. The collection of color transparencies of fungi of South Carolina that belonged to the late Col. E. R. Davies, Charleston, South Carolina, contains several pictures of what appear to be brown tinted specimens of *A. flavoconia*. Tulloss has examined exsiccati sent from XAL by G. Guzmán (*Garcia 351* and *Guzmán 24422*); in these cases the collecting notes were very limited, but suggested *A. flavoconia*; the Q' for 60 spores from three specimens was 1.21.

The immature specimens of *Franco-M. 420* (CUVC, NY) appear to belong in this taxon.

Henao-M. and Tobón-A. (1984) included three specimens in *A. flavoconia*; it is probable that these are referable to var. *inquinata* or to the following variety.

⟶

Figs. 21–28. Microscopic features. Figs. 21–23. *Amanita brunneolocularis*. 21. Hymenium and subhymenium (*Ovrebo 2506*). 22. Universal veil from pileus (*Halling 5247*). 23. Partial veil (*Ovrebo 2506*). Figs. 24–26. *Amanita flavoconia* var. *inquinata* (*Halling 5067*). 24. Hymenium and subhymenium. 25. Partial veil. 26. Universal veil from pileus. Figs. 27–28. *Amanita picea* (*Halling 5250*). 27. Hymenium and subhymenium. 28. Universal veil from pileus; interior of volval wart. Scale line = 10 μm.

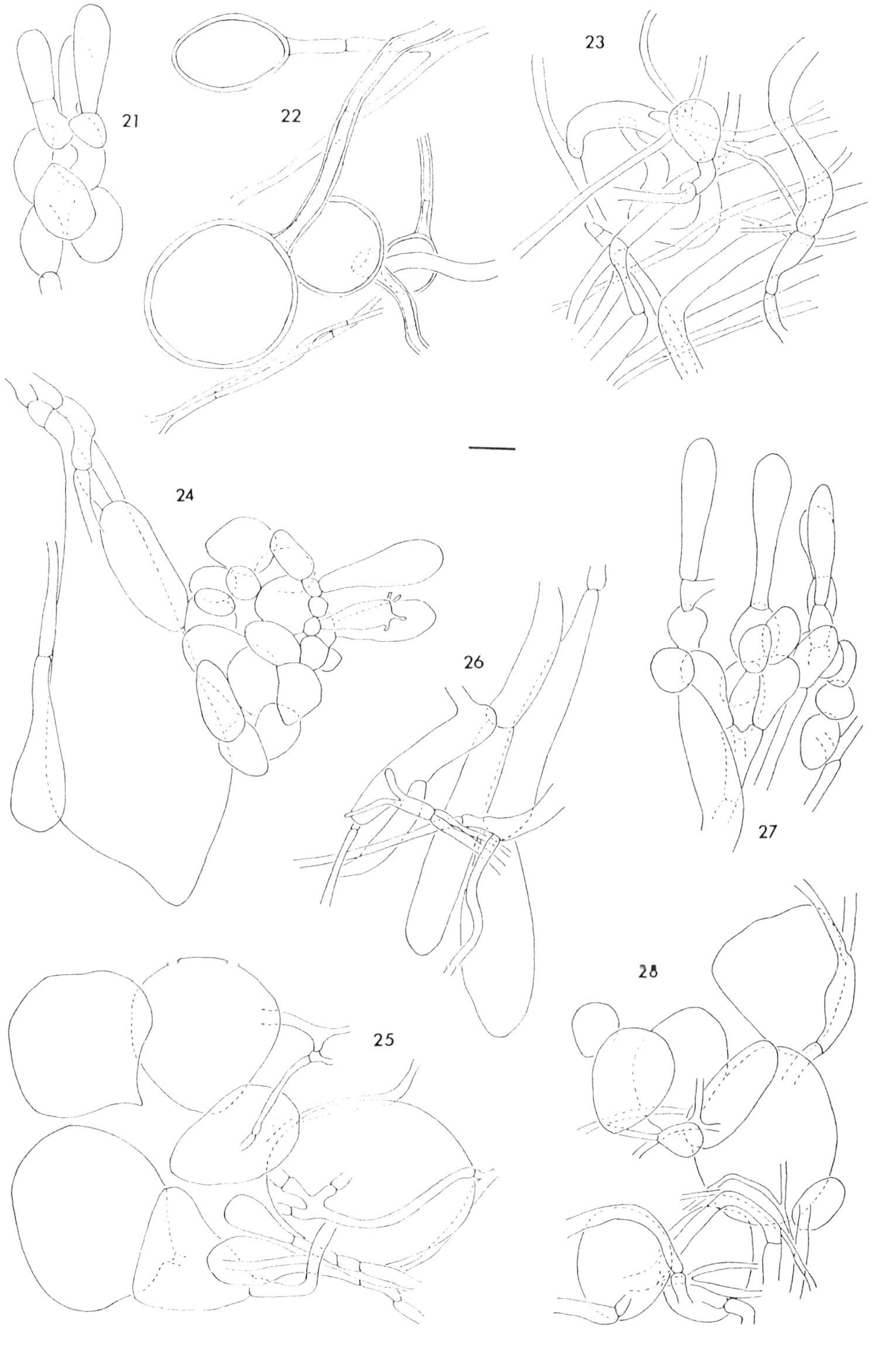
21
22
23
24
26
27
28
25

12. **Amanita flavoconia** Atk. var. **sinapicolor** Tulloss, Ovrebo & Halling, *var. nov.*

Typus: COSTA RICA. SAN JOSÉ: 5 km S of Empalme, Jardín de Dota, Jul 1983, *Gomez & Alfaro 20616* (holotypus: F 1073280).

Etymology: *sinapis*, mustard + *color*.

A typo pileo sinapiflavo et sporis minoribus ratione 'Q' *humili differt.*

PILEUS: 28–37 mm diam., mustard yellow to dull yellow; margin nonstriate or slightly striate when mature; universal veil remnants in scattered yellow patches. LAMELLAE: white, sometimes yellow near pileus margin. STIPE: 43–58 × 3–4 mm, white with yellow regions, narrowed upward; annulus superior to submedian, persistent, membranous, yellow; universal veil in small yellow patches easily lost in soil. Odor yeasty. Taste not recorded.

PILEIPELLIS: 35–80 μm thick, with upper portion hyaline and lower portion orangy-yellow; filamentous, undifferentiated hyphae 2.2–4 μm diam., partially gelatinized, subradially arranged; oleiferous hyphae uncommon, 3.5–5 μm diam. PILEUS CONTEXT: filamentous, undifferentiated hyphae 1.8–8.5 μm diam., branched; inflated cells ellipsoid to elongate to ventricose, to 90 × 36 μm; oleiferous hyphae scarce, ±3.2 μm diam. LAMELLA TRAMA: bilateral, with shallow angle of divergence, narrow; central stratum distinct, relatively narrow; filamentous, undifferentiated hyphae 1–6.8 μm diam.; inflated cells terminal or as chain of inflated hyphal segments, to 135 × 38 μm, subglobose to irregularly subpyriform to ellipsoid to ovoid to elongate to subclavate to clavate, thin-walled; oleiferous hyphae uncommon, ±2 μm diam. SUBHYMENIUM: cellular; basidia arising from a thin layer of small inflated cells and these arising from a thicker layer of larger inflated cells. BASIDIA: 23–37 × 6.5–11.5 μm, dominantly 4-, occasionally 2- or 1-sterigmate, clavate, thin-walled; clamps not observed. UNIVERSAL VEIL: Absent from pileus. On stipe base: filamentous, undifferentiated hyphae 1.8–10 μm diam., frequently branched, plentiful, irregularly arranged, with terminal cell often slightly inflated (to narrowly clavate); inflated cells thin-walled, subglobose to ellipsoid to elongate, to 92 × 69 μm, plentiful, sometimes with yellow tint, terminal, sometimes in short chains (to three cells); oleiferous hyphae common, 1.8–4 μm diam. STIPE CONTEXT: acrophysalidic; filamentous, undifferentiated hyphae 1.2–8.5 μm diam., sparsely branched, plentiful; acrophysalides dominant, thin-walled, to 200+ × 50 μm; oleiferous hyphae uncommon, 3.5–17.5 μm diam. PARTIAL VEIL: filamentous, undifferentiated hyphae dominant, 1.5–10 μm diam., branched, with a significant number subradially arranged, but also with many hyphae loosely to densely tangled and interwoven; inflated cells terminal or in chains, smallest ones ovoid, but majority allantoid, to 171 × 23 μm; oleiferous hyphae scarce to absent.

BASIDIOSPORES: [134/7/5] (6–)6.2–8(–9) × (4.8–)5.8–7(–8.2) μm [**L** = 6.6–7.5 μm; **L'** = 7.1 μm; **W** = 6–6.5 μm; **W'** = 6.3 μm; Q = (1–)1.03–1.25(–1.37); **Q** = 1.10–1.16; **Q'** = 1.13], amyloid, hyaline, thin-walled, smooth, globose to subglobose to broadly ellipsoid, occasionally ellipsoid; contents guttulate; apiculus sublateral, small, cylindric; color in deposit not recorded.

Distribution and habitat: Subgregarious, under *Quercus humboldtii*, sometimes mixed with *Pinus patula* Schiede & Deppe *ex* Schlect. and *Eucalyptus* sp. at 1900–2850 m elevation in Colombia (March to May) and under *Quercus* at 1800–2000 m elevation in Costa Rica (June to July). In one case (*Mueller 2819*), growing with *A. flavoconia* var. *inquinata*.

Collections examined: COLOMBIA. ANTIOQUIA: Mpio. Guarne, 14 km E of Medellín, Centro Experimental Piedras Blancas, 15 May 1987, *Halling, Mueller & Strack* [*GMM 2850*] (F 1080775); Mpio. Santa Rosa de Osos, vereda "El Chaquiro," road from Santa Rosa de Osos to Aragón, near Llanos de Cuivá, 13 May 1987, *Halling, Mueller & Strack* [*GMM 2819*] (F 1080754, mixed collection). CUNDINAMARCA: Between Pacho and Páramo Guerrero, 9 Apr 1968, *Singer B6029* (F 1014373).

Extralimital collections examined: COSTA RICA. CARTAGO: Along Hwy. 2, Cerro de la Muerte, 22 Jun 1986, *Ovrebo 2004* (CSU).

DISCUSSION

Amanita flavoconia var. *sinapicolor* (unillustrated) differs from both var. *flavoconia* and var. *inquinata* in having a mustard yellow to sordid yellow pileus and smaller spores with a lower range for Q. We consider pileus color to be the primary differentiating character between var. *sinapicolor* and var. *inquinata*. 'F' test results indicate that L' and Q' are significantly different in these two varieties, but spore size and shape alone may not be sufficient to diagnose an individual specimen that lacks annotation concerning its pileus color when fresh. Also, see the discussion under *A. flavoconia* var. *inquinata*, above. Tissues of var. *sinapicolor* may contain fewer oleiferous hyphae than corresponding tissues in var. *inquinata*; however, the presence of such hyphae is a somewhat variable character in *Amanita*.

13. **Amanita picea** Tulloss, Ovrebo & Halling, *sp. nov.* Figs. 27–28, 36

Typus: COLOMBIA. BOYACÁ: Mpio. San Miguel de Sema, road from Simijacta to San Miguel de Sema, 9 May 1987, *Halling 5250* (holotypus: COL; isotypus: NY).

Etymology: *piceus*, pitch-black, black changing to brown.

Pileus 23–105 mm latus, piceus; margo estriatus, exappendiculatus; materiae volvicae nullae aut ex assumentis aut verrucis fibrillosis vel gossypinis griseis vel atro-brunneo-griseis pagina inferiori pallida constantes. Stipes 40–100 × 10–30 mm, pallidus, fibrillis griseis; caro alba, decolorationibus denique ochraceis; annulus subapicalis, membranaceus, desuper albus, subter subgriseus; materiae volvicae circa apicem bulborum cingulos 2–3 verrucosos pyramidales ad subpyramidales griseos formantes. Fibulae nullae. Sporae amyloideae, subglobosae vel late ellipsoideae vel ellipsoideae vel interdum elongatae, (7.5–)8.8–11.2(–12.5) × (5.5–)6.5–8.2(–9) μm.

PILEUS: 23–105 mm diam., black at first then dark brown (almost black) at center to 6E7 or paler toward margin, becoming shiny when dry, ovoid to hemispheric at first then obtusely convex expanding to plano-convex, becoming virgate in age; context white, ±4 mm thick, unchanging when cut; margin not striate, nonappendiculate, incurved to decurved; universal veil absent or in detersile, appressed, submembranous, fibrillose to cottony, gray to dark brownish gray patches with pallid undersides and margins or in detersile warts that are dark brownish gray or (pinkish) gray with darker gray centers and with pallid margins, in all cases developing rusty or fulvous spots. LAMELLAE: close to crowded, free to very narrowly adnate, white, becoming pale orangish cream or darker (5A3 to 5A5 or 7.5YR 7–8/6) when dry, ±5 mm broad; margin minutely fimbriate (with 10× lens); lamellulae truncate. STIPE: 40–100 × 10–30 mm, pallid with a dense layer of longitudinally arranged gray fibrils, white when young, developing pinkish orange stains with age and near base with handling or as dark as 5C–D4 at maturity, subcylindrical to subventricose; annulus subapical to superior, membranous, pendant, flaring, persistent, white above, pale gray below, sometimes striate above; bulb subradicating, splitting longitudinally at broadest point; a patch of submembranous, gray universal veil material (limbus internus) found on the lower stipe of one specimen; context white; universal veil gray, forming two to three concentric rings (around the top of the bulb) of coarse, floccose, verruculose, minutely pyramidal to subpyramidal warts with dark peaks.

PILEIPELLIS: 225–290 μm thick, some brown pigment dissolving in mount (wetted with ethanol, mounted in 2% KOH), extensively gelatinized; filamentous, undifferentiated hyphae 1.8–6 μm diam., branched, interwoven, arranged disorderly at surface, below surface subradially arranged, many with dark brown intracellular pigment scattered throughout except for hyphae nearest surface and nearest transition to pileus context; oleiferous hyphae 1–6 μm diam. PILEUS CONTEXT:

filamentous, undifferentiated hyphae 3.2–21.5 μm diam., branched, tangled, occasionally coiled, loosely interwoven in open lattice; inflated cells clavate to elongate to ellipsoid, to 138 × 80 μm, terminal or in chains and then not so fully inflated, thin-walled or with walls very slightly thickened; oleiferous hyphae relatively common, 5.8–12 μm diam. LAMELLA TRAMA: bilateral; central stratum narrow, distinct; angle of divergence of inflated cells mostly 45°–60°; filamentous, undifferentiated hyphae 2–8.5 μm diam., branched; inflated cells ellipsoid to ovoid to elongate, to 74 × 35 μm, thin-walled; oleiferous hyphae 1.5–4.5 μm diam. SUBHYMENIUM: cellular, 5–7 cells deep; basidia arising from small ovoid to subglobose to clavate cells in branched chains or from short inflated or uninflated hyphal segments. BASIDIA: 37–55 × 9–12.2 μm, 4- or infrequently 2-sterigmate, thin-walled; clamps not observed. UNIVERSAL VEIL: On pileus: filamentous, undifferentiated hyphae 1–14 μm diam., branched, plentiful in upper surface of warts; inflated cells dominant throughout warts except at upper surface, gelatinized at interface with pileipellis, hyaline or with pale grayish brown intracellular pigment, elongate-ventricose to ellipsoid to ovoid to subglobose, to 56 × 47 μm, thin-walled, terminal, single or in short chains; branched, oleiferous hyphae present, 2.2–12 μm diam., locally plentiful especially near upper surface. On stipe base: very similar to that on pileus with the following differences: somewhat more gelatinized and with a somewhat larger proportion of filamentous, undifferentiated hyphae which are more commonly in fascicles; inflated cells somewhat larger, to 73 × 53 μm; oleiferous hyphae uncommon, ±1.5 μm diam. STIPE CONTEXT: acrophysalidic; filamentous, undifferentiated hyphae 1.8–11.5 μm diam., branched, plentiful, often in fascicles of those of smaller diameter; acrophysalides to 271 × 125 μm, most narrower, with walls thin or very slightly thickened; oleiferous hyphae 1.5–6.5 μm diam., sometimes appearing intermixed in fascicles of filamentous, undifferentiated hyphae, occasionally forming fascicles nearly entirely of oleiferous elements. PARTIAL VEIL: almost entirely of filamentous, undifferentiated hyphae 2.5–9.5 μm diam., interwoven, branched, in large part subradially arranged, sometimes in fascicles; inflated cells scarce, terminal, clavate to broadly clavate, to 59 × 37 μm; oleiferous hyphae locally plentiful in knots and tangles, 2–11 μm diam.

BASIDIOSPORES: [127/6/2] (7.5–)8.8–11.2(–12.5) × (5.5–)6.5–8.2(–9) μm [**L** = 9.3–10.6 μm; **L'** = 10.1 μm; **W** = 6.9–7.5 μm; **W'** = 7.2 μm; Q = (1.11–)1.28–1.60(–1.74); **Q** = 1.35–1.47; **Q'** = 1.40], amyloid or slightly so, hyaline, thin-walled or sometimes with walls thickened in the apical region, smooth, subglobose to broadly ellipsoid to ellipsoid, occasionally elongate, adaxially flattened or slightly so; contents guttulate; apiculus sublateral, cylindric, rather small; color in deposit not recorded.

Distribution and habitat: Gregarious to subgregarious under *Quercus humboldtii* in loamy soil to 2700 m elevation, May and November.

Collection examined: COLOMBIA. NARIÑO: Mpio. Pasto, vereda "La Josefina," Km 17, road from Pasto to Chachagüí, 20 Nov 1988, *Franco-M. 164* (NY, PSO).

Discussion

Amanita picea (Fig. 36) is a medium-sized mushroom with a shiny black pileus and plentiful pyramidal to subpyramidal gray warts in circles on the base of the stipe and, if not lost, irregularly dispersed or confluent as patches on the pileus. In the field, this unique and easily recognizable species looks like a black form of *A. muscaria*, but it can be separated easily from section *Amanita* by its amyloid spores.

Recent lists of *Amanita* species known from South America are available (*e.g.*, Raithelhuber, 1986; Garrido & Bresinsky, 1985; Garrido, 1987). In these sources we find no taxon reported in section *Validae* with a spore length/width ratio similar to that computed for *A. picea*. The world literature contains descriptions of only four taxa in this section with similar spore size, although the values of Q determined for *A. picea* are very commonly found in species of section *Validae*. The four taxa with similar spores are macroscopically quite dissimilar from *A. picea*:

—*Amanita bharatensis* Sathe & David (Sathe *et al.*, 1980) of southwestern India has a bright red pileus, white universal veil material, slightly larger spores and may belong in section *Amanita*; the protologue strongly suggests *A. muscaria* var. *muscaria* except for the report of faintly amyloid spores.

—*Amanita flavorubescens* Atk. of eastern North America usually has yellow tones in the pileus and a bright yellow universal veil that does not break up into subpyramidal warts. The flesh in the stipe base and bulb slowly stain wine-red in this species.

—*Amanita griselloides* Reid (1980) of Australia has a silvery gray pileus and a complex universal veil; it may belong in section *Lepidella*.

—*Amanita rubescens* has different pileus coloration and develops red-brown staining throughout the basidiocarp. It is discussed above in comparison to *A. brunneolocularis*.

Amanita morrisii Peck of eastern North America is somewhat similar to *A. picea* in the colors of its pileus in early stages of expansion. Tulloss has seen unpublished watercolors of topotypes made by the original collector (G. E. Morris) and now located in PM and NYS; these paintings show a pink underside to the annulus, a nearly naked bulb, and very pale submembranous (not wart-like) patches of universal veil material. Tulloss (1991) reported the spores of *A. morrisii* to be [485/24/6] (6–) 7.2–9.5(–11.5) × (4.2–)5.5–7(–8.2) μm with Q = (1.28–)1.30–1.39(–1.42).

Acknowledgments

We appreciate the careful collecting efforts, field notes, photography, translation of Spanish notes and text, and other supportive assistance of A. E. Franco-Molano (NY); she is also responsible for making the phenoloxidase spot test observations which we report. We are grateful for assistance in Colombia from the following: Dr. R. Callejas, Dra. L. Atehórtua, Prof. Y. Saldarriaga and collaborators, Prof. O. Salazar, Prof. J. Ramos, Dr. P. Silverstone-Sopkin, Ms. Z. Piñeros, Ms. R. Ruíz, Ms. B. Echeverry and Ms. P. Velásquez. We express our sincere gratitude to the following: Dr. Rupert Barneby (NY) for his assistance in preparing the Latin diagnoses and in formation of specific epithets; Dr. Cornelis Bas (L) for providing Tulloss working space in his laboratory, arranging for the loans of relevant material, and reviewing this paper; Drs. D. E. Desjardin (SFSU) and H. D. Thiers (SFSU) for their valuable reviews; Mr. Joaquín Cifuentes Blanco, for the loan of material from FCME; Dr. Gaston Guzmán and Mr. Victor M. Bandala for the loan of material from XAL; Dr. R. P. Korf (CUP), Dr. D. H. Pfister (FH), Dr. D. N. Pegler (K), Dr. J. Haines (NYS), and Dr. P. Fechko (PM) for the loan of, or access to, pertinent specimens; Drs. Gregory M. Mueller and Rolf Singer for their support in providing loans of material from F and for access to Dr. Singer's notes on *Amanita* spp.; Dr. Ronald H. Petersen (TENN) for assistance in obtaining loans from Mexican herbaria; and Dr. Jorge E. Wright for the loan of essential material from BAFC; Ms. Mary A. King, Roosevelt, New Jersey for assisting in preparation of the manuscript for publication. Support from National Science Foundation Grant #8600424 to Halling made possible much of the collecting that underlies this paper. Field work in Costa Rica by Ovrebo was supported by a grant from the Tinker Foundation administered by the Mesoamerican Ecology Institute, Tulane University.

Literature Cited

Aroche, R. M., M. Villegas, J. Cifuentes, F. Lorea & J. Bonavides. 1984. Nuevos datos sobre la distribución y taxonomía de *Amanita phalloides* en México. Bol. Soc. Mex. Micol. **19:** 275–281.

Atkinson, G. F. 1914. The development of *Amanitopsis vaginata*. Ann. Mycol. **12:** 369–392.

Bas, C. 1969. Morphology and subdivision of *Amanita* and a monograph of its section *Lepidella*. Persoonia **5:** 285–579.

_______. 1974. A rare but widespread *Amanita* associated with *Alnus*. Travaux mycologiques dédiés à R. Kühner. Bull. Soc. Linn. Lyon, Num. Spéc.: 17–23.

_______. 1978. Studies in *Amanita*—some species from Amazonia. Persoonia **10:** 1–22.

_______. 1984. On the correct name of '*Amanita inaurata* Secr.' Persoonia **12:** 192–193.

Bertault, R. 1965. Amanites du Maroc (deuxième contribution). Bull. Soc. Mycol. France **81:** 345–371.

Boedijn, K. B. 1951. Notes on Indonesian fungi. The genus *Amanita*. Sydowia **5:** 317–327.

Corner, E. J. H. & C. Bas. 1962. The genus *Amanita* in Singapore and Malaya. Persoonia **2:** 241–304.

Forman, L. L. 1963. *Trigonobalanus*, a new genus of Fagaceae, with notes on the classification of the family. Kew Bull. **17:** 381-396.

Garrido, N. 1987. Agaricales *s.l.* und ihre Mykorrhizen in den *Nothofagus*-Wäldern Mittelchiles. Doctoral dissertation. Regensburg Univ. pagination unknown.

_______ **& A. Bresinsky.** 1985. *Amanita merxmuelleri* (Agaricales), eine neue Art aus *Nothofagus*-Wäldern Chiles. Bot. Jahrb. Syst. **107:** 521–540.

Gilbert, J.-E. 1918. Le genre *Amanita* Persoon. Lons-Le-Saunier. 186 pp.

_______. 1940–1941. Amanitaceae. Iconographia Mycologica **26:** 1–427.

Guzmán, G. 1982. New species of fungi from the Yucatán Peninsula. Mycotaxon **16:** 249–261.

_______ **& L. Varela.** 1978. Los hongos de Colombia III. Observaciones sobre los hongos, líquenes y mixomycetos de Colombia. Caldasia **12:** 309–338.

Halling, R. E. 1989a. A synopsis of Colombian boletes. Mycotaxon **34:** 93–113.

_______. 1989b [1990]. Notes on *Collybia* III. Three neotropical species of subg. *Rhodocollybia*. Mycologia **81:** 870–875.

_______ **& C. L. Ovrebo.** 1987a. A new species of *Rozites* from oak forests of Colombia, with notes on biogeography. Mycologia **79:** 674–678.

_______ **&** _______. 1987b. Some agarics and boletes of Colombian oak forests. Mycol. Soc. Amer. Newslett. **38(1):** 27 [Abstract].

Heinemann, P. 1964. Les amanitées. Les Naturalistes Belges, Bruxelles. 22 pp.

Henao-M., L. G. & L. E. Tobón-A. 1984. Macromicetos de la cuenca de la quebrada Piedras Blancas (Departamento de Antioquia). Bachelor's Thesis, Universidad Nacional, Medellín. 237 pp.

Herter, W. G. 1934. La aparición del hongo venenoso *Amanita phalloides* en Sudamérica. Revista Sudamer. Bot. **1:** 111–119.

Holmgren, P. K., N. H. Holmgren & L. C. Barnett. 1990. Index herbariorum. Part I. The herbaria of the world. Ed. 8. New York Botanical Garden, Bronx, NY. 693 pp.

Imai, S. 1933. Studies on the Agaricaceae of Japan. I. Volvate agarics in Hokkaido. Bot. Mag. (Tokyo) **47:** 423–432.

_______. 1938. Studies of the Agaricaceae of Hokkaido. I. J. Fac. Agric. Hokkaido Univ. **43:** 1–178.

Imazeki, R. & T. Hongo. 1987. Colored illustrations of Mushrooms of Japan. Vol. I. Hoikusha, Osaka. 325 pp.

_______, **Y. Otani & T. Hongo.** 1988. Fungi of Japan. Yama-Kei, Tokyo. 624 pp.

Jenkins, D. T. 1977. A taxonomic and nomenclatural study of the genus *Amanita* section *Amanita* for North America. Biblio. Mycol. **57:** 1–126.

_______. 1978. A study of *Amanita* types I. Taxa described by C. H. Peck. Mycotaxon **7:** 23–44.

_______. 1986. *Amanita* of North America. Mad River Press, Eureka. 198 pp.

_______. 1988. A new species of *Amanita* from North America: *Amanita levistriata*. Mycotaxon **32:** 415–419.

Kauffman, C. H. 1918. The Agaricaceae of Michigan. I. Wynkoop Hallenbeck Crawford, Lansing. 924 pp.

Kelley, K. L. 1965. Color name charts illustrated with centroid colors. Standard sample #2106 supplement to National Bureau of Standards Circular 553. U. S. Government Printing Office, Washington, D.C. unpaginated.

Knudsen, H. & T. Borgen. 1987. Agaricaceae, Amanitaceae, Boletaceae, Gomphidiaceae, Paxillaceae and Pluteaceae in Greenland. *In*: G. A. Laursen, J. F. Ammirati & S. A. Redhead (eds.), Arctic and Alpine Mycology **2:** 235–253. Plenum, New York.

Kornerup, A. & J. H. Wanscher. 1983. Methuen handbook of colour. Ed. 3. Eyre Methuen, Ltd., London. 252 pp.

Krieger, L. C. C. 1935. The mushroom handbook. MacMillan and Co. Reprinted in 1967 by Dover, New York. 560 pp.

Kühner, R. & H. Romagnesi. 1974. Flore analytique des champignons supérieurs. Masson et Cie., Paris. 2nd printing. 557 pp.

Lange, M. & F. B. Hora. 1963. Collins Guide to Mushrooms and Toadstools. Collins, London. 257 pp.

Maerz, A. & M. R. Paul. 1930. A dictionary of color. McGraw-Hill, New York. 22 pp. + 207 pl.

Marchand, A. 1971. Champignons du nord et du midi. Vol. 1. Hachette, Perpignan. 282 pp.

_______.1973. Champignons du nord et du midi. Vol. 2. Hachette, Perpignan. 273 pp.

Marr, C. D. 1979. Laccase and tyrosinase oxidation of spot test reagents. Mycotaxon **9:** 244–276.

Massart, F. 1984. Approche du genre *Amanita*. Société Linnéenne de Bordeaux. 138 pp.

_______ **& P. Beauvais.** 1975. Une variété ecologique nouvelle d'*Amanita phalloides*. Bull. Hist. Nat. Soc. Linn. Bordeaux **5:** 9–12.

Merlo, E. G. & M. Traverso. 1983. Le amanite. Sagep, Genoa. 151 pp.

Moser, M. 1983. Keys to agarics and boleti. Translated from German by S. Plant. Phillips, London. 535 pp.

Mueller, G. M. 1991. The Swedish taxa of *Laccaria* (Tricholomataceae) with notes on their distribution. Nord. J. Bot. **10:** 665–680.

Munsell Color. 1975. Munsell soil color charts. Baltimore, Maryland. unpaginated.

Nasi, M. 1977. Los hongos superiores de la Sabana de Bogotá y alrededores. Master's Thesis, Universidad de los Andes, Bogotá. pagination unknown. [We have not been able to locate a copy of this work, but include it here for purposes of completeness.]

Parrot, A. G. 1960. Amanites du sud-ouest de la France. Centre d'Études et de Recherches Scientifique, Biarritz. 173 pp.

Peck, C. H. 1879 [1880]. Report of the botanist. Annual Rep. New York State Mus. **33:** 11–49.

Pegler, D. N. 1983. Agaric flora of the Lesser Antilles. Kew Bull. Add. Ser. **9:** 1–615.

Phillips, R. 1981. Mushrooms and other fungi of Great Britain and Europe. Pan, London. 288 pp.

Pomerleau, R. 1980. Flore des champignons au Québec. La Presse, Montréal. 653 pp.

Pulido-Osuna, M. M. 1983. Estudios en Agaricales colombianos—Los hongos de Colombia IX. Instituto de Ciencias Naturales, Museo de Historia Natural, Biblioteca José Jerónimo Triana, 7. Editorial Presencia, Bogotá. 143 pp.

Quélet, L. 1888. Flore mycologique de la France et des pays limitrophe. Doin, Paris. 492 pp.

Raithelhuber, J. 1986. Amanitaceae in Südamerika. Metrodiana **14:** 3–21.

________. 1987. Flora Mycologica Argentina. Hongos I. Stuttgart. 405 pp.

Reid, D. A. 1980. A monograph of the Australian species of *Amanita* Pers. ex Hook. (Fungi). Austral. J. Bot. Suppl. Ser. **8:** 1–96.

Santiago, G., J. Cifuentes & M. Villegas. 1984. Contribución al conocimiento del género *Amanita* subgénero *Amanita* en México. Bol. Soc. Mex. Micol. **19:** 93–105.

Sathe, A. V., S. Deshpande, S. M. Kulkarni & J. Daniel. 1980. Agaricales (mushrooms) of southwest India. Maharashtra Association for the Cultivation of Science Research Institute, Pune. 114 pp.

Schaeffer, J. C. 1774. Fungorum qui in Bavaria et Palatinatu circa Ratisbona nascuntur icones nativis coloribus expressae **4:** 1–136.

Singer, R. 1954. Agaricales von Nahuel Huapí. Sydowia **8:** 100–157.

________. 1959. Basidiomycetes from Masatierra. Ark. Bot. II, **4:** 371–400.

________. 1963. Oak mycorrhiza fungi in Colombia. Mycopathol. Mycol. Appl. **20:** 239–252.

________ **& M. Moser.** 1965. Forest mycology and forest communities in South America. Mycopathol. Mycol. Appl. **26:** 130–191.

________, **C. L. Ovrebo & R. E. Halling.** 1990. New species of *Phylloporus* and *Tricholomopsis* from Colombia, with notes on *Phylloporus boletinoides*. Mycologia **82:** 452–459.

Smith, A. H., H. V. Smith & N. S. Weber. 1979. How to know the gilled mushrooms. Wm. C. Brown, Dubuque. 334 pp.

Thiers, H. D. & J. F. Ammirati, Jr. 1982. New species of *Amanita* from western North America. Mycotaxon **15:** 155–166.

Tulloss, R. E. 1990. *Amanita crenulata*—history, taxonomy, distribution, and poisonings. Mycotaxon **39:** 393–405.

_______. 1991. *Amanita morrisii*—history, taxonomy, and distribution. Mycotaxon **40**: 281–286.

_______ **& D. T. Jenkins.** 1986. Notes on distribution of *Amanita albocreata*. Mycotaxon **26**: 81–83.

_______ **& G. Wright.** 1988. *Amanita protecta*—a new species from coastal southern California. Mycotaxon **34**: 615–622.

Vergeer, P. P. 1986. The Meixner test evaluated. McIlvainea 7(**2**): 61–68.

Figs. 29, 30. Basidiomata of *Amanita xylinivolva* and *A. fuligineodisca*. **29.** *A. xylinivolva* (*Ovrebo 2487*) (×¾). **30.** *A. fuligineodisca* (*Ovrebo 2482*) (×¾).

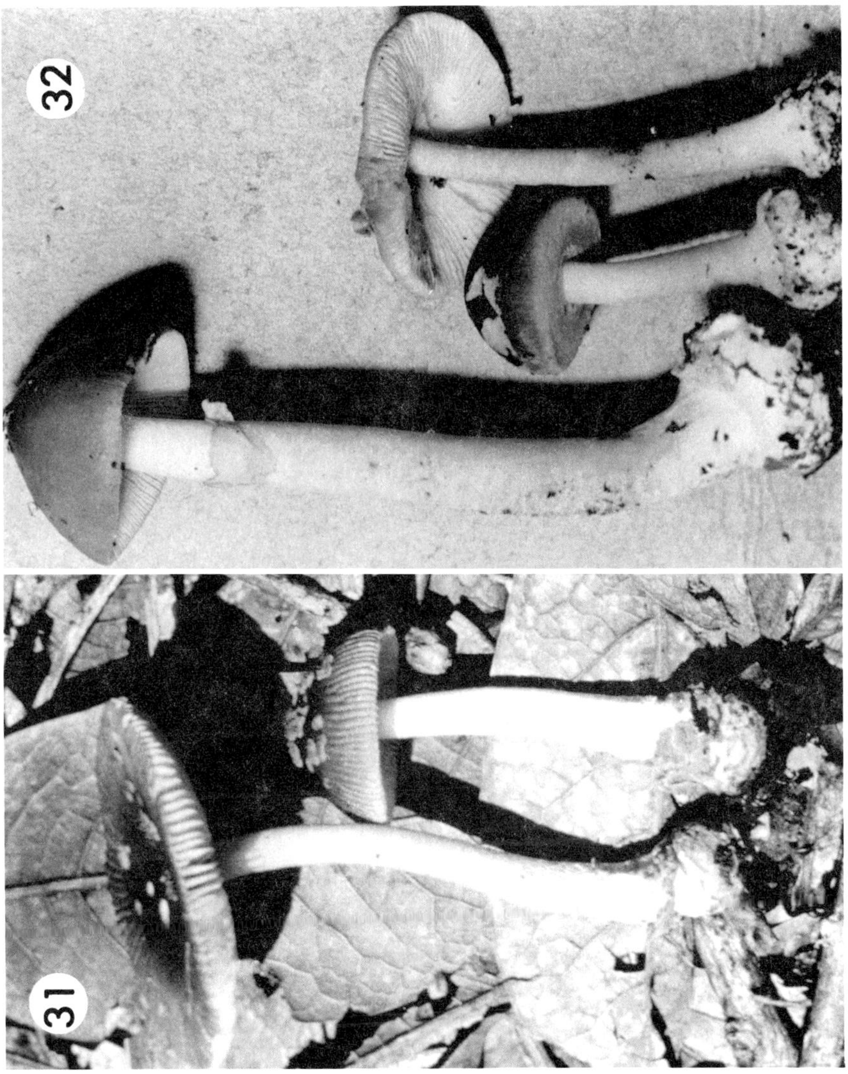

Figs. 31, 32. Basidiomata of *Amanita colombiana* and *A. flavoconia* var. *inquinata*. **31.** *A. colombiana* (*Ovrebo 2425*) (×1). **32.** *A. flavoconia* var. *inquinata* (*Halling 5067*) (×1¾).

Figs. 33, 34. Basidiomata of *Amanita advena* and *A. arocheae*. **33.** *A. advena* (*Ovrebo 2547*) (×1). **34.** *A. arocheae* (*Halling 5071*) (×1).

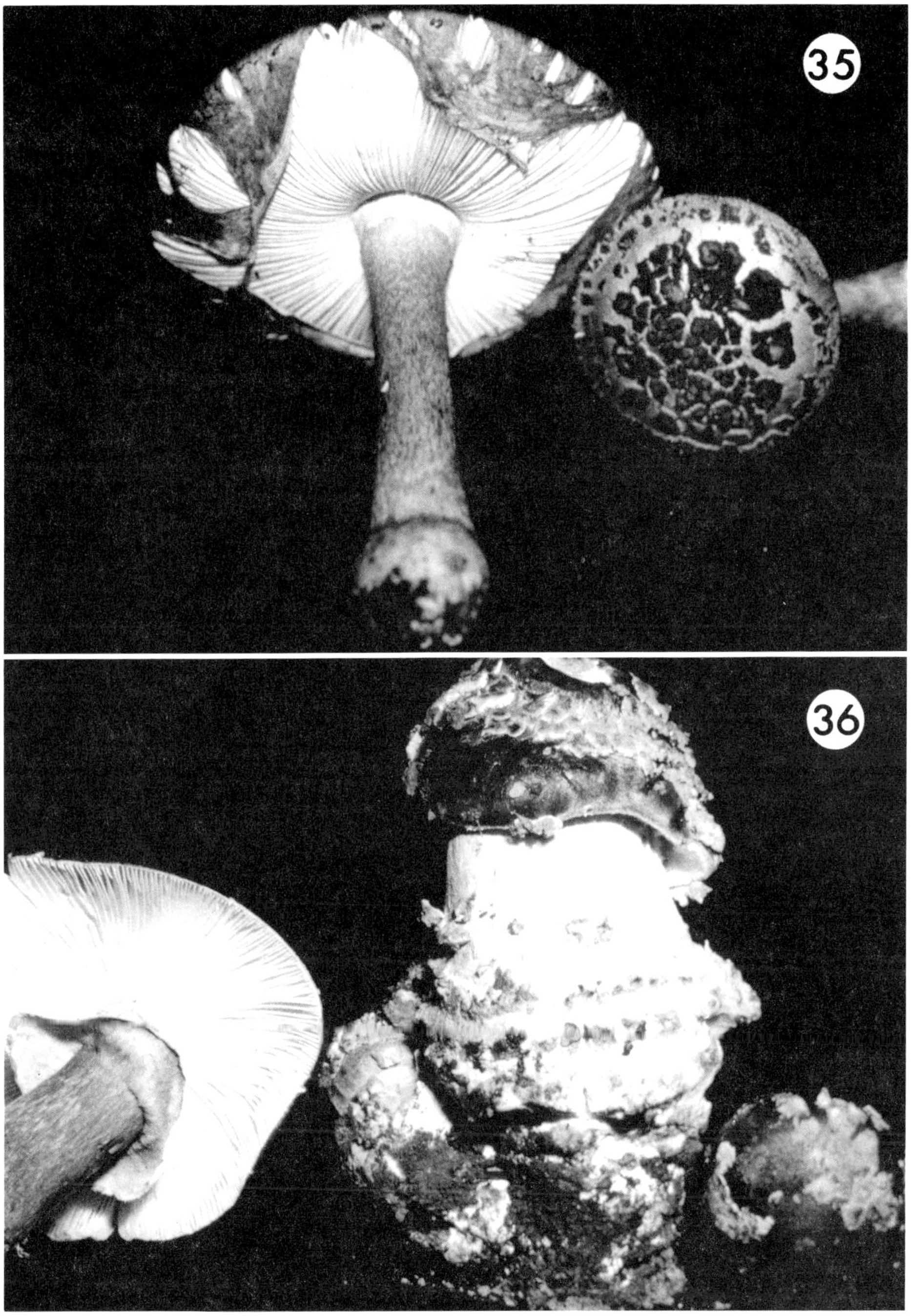

Figs. 35, 36. Basidiomata of *Amanita brunneolocularis* and *A. picea*. **35.** *A. brunneolocularis* (*Ovrebo 2506*) (×1¾). **36.** *A. picea* (*Halling 5250*) (×1).

Index to Taxa

Boldface page numbers with * indicate pages with descriptions.

MEMOIRS OF THE NEW YORK BOTANICAL GARDEN

The Botany of the Guayana Highland—Parts I-XIII
by Bassett Maguire and Collaborators

I.	Mem. N. Y. Bot. Gard. 8(2): 87–160. 1953. (Combretaceae, Compositae, Eriocaulaceae, Leguminosae, Malpighiaceae, Melastomataceae, Polygalaceae, Rutaceae)	$10.00
II.	Mem. N. Y. Bot. Gard. 9(3): 235–392. 1957. (Annonaceae, Begoniaceae, Bignoniaceae, Bromeliaceae, Compositae, Droseraceae, Eriocaulaceae, Gramineae, Haemadoraceae, Leguminosae, Musci, Piperaceae, Polygalaceae)	$15.00
III.	Mem. N. Y. Bot. Gard. 10(1): 1–156. 1958. (Apocynaceae, Bignoniaceae, Combretaceae, Guttiferae, Leguminosae, Melastomataceae, Myrtaceae, Rapateaceae, Xyridaceae)	$15.00
IV.	Mem. N. Y. Bot. Gard. 10(2): 1–37. 1960. (Musci)	$10.00
IV.	Mem. N. Y. Bot. Gard. 10(4): 1–87. 1960. (Acanthaceae, Annonaceae, Aquifoliaceae, Bromeliaceae, Guttiferae, Linaceae, Marantaceae, Melastomataceae, Ochnaceae, Rutaceae, Xyridaceae)	$10.00
V.	Mem. N. Y. Bot. Gard. 10(5): 1–278. 1960. (Bambusoideae, Bromeliaceae, Guttiferae, Lauraceae, Melastomataceae, Piperaceae, Rubiaceae, Xyridaceae)	$17.50
VI.	Mem. N. Y. Bot. Gard. 12(3): 1–285. 1965. (Aquifoliaceae, Cyperaceae, Euphorbiaceae, Gramineae, Lauraceae, Rapateaceae, Rubiaceae)	$17.50
VII.	Mem. N. Y. Bot. Gard. 17(1): 1–439. 1967. (Abolbodaceae, Bombacaceae, Compositae, Cyperaceae, Euphorbiaceae, Labiatae, Lentibulariaceae, Polypodiaceae, Rubiaceae)	$35.00
VIII.	Mem. N. Y. Bot. Gard. 18(2): 1–290. 1969. (Annonaceae, Apocynaceae, Bromeliaceae, Cyperaceae, Gramineae, Malpighiaceae, Melastomataceae, Myrtaceae, Quiinaceae, Schizaeaceae, Velloziaceae)	$20.00
IX.	Mem. N. Y. Bot. Gard. 23: 1–832. 1972. (Bonnetiaceae, Caryocaraceae, Clusiaceae, Cycadeaceae, Cyclanthaceae, Gramineae, Meliosmaceae, Melastomataceae, Palmae, Piperaceae, Polygalaceae, Polypodiaceae, Rubiaceae, Sapotaceae, Tetrameristaceae)	O/P
X.	Mem. N. Y. Bot. Gard. 29: 1–288. 1978. (Bonnetiaceae, Bignoniaceae, Cyatheaceae, Ericaceae, Icacinaceae, Saccifoliaceae, Sarraceniaceae, Styracaceae, Symplocaceae, Tiliaceae)	$25.00
XI.	Mem. N. Y. Bot. Gard. 32: 1–391. 1981. (Tepuianthaceae, Malpighiaceae, Dipterocarpaceae, Convolvulaceae, Ebenaceae, Lissocarpaceae, Gentianaceae)	$40.00
XII.	Mem. N. Y. Bot. Gard. 38: 1–84. 1984. (Hymenophyllopsidaceae, Hymenophyllaceae, Araliaceae)	$16.00
XIII.	Mem. N. Y. Bot. Gard. 51: 1–127. 1989. (Gentianaceae, *Ouratea* [Ochnaceae], Melastomataceae, Dioscoreaceae, Nyctaginaceae, Monimiaceae, Capparaceae, Cunoniaceae, Rhamnaceae, Rutaceae)	$28.75

Postage and Handling:

U.S. Orders:	$3.00 + 4% of subtotal.
Non-U.S. Orders:	$4.00 + 6% of subtotal. Payment in U.S. currency drawn on a U.S. bank. Thank you.

Order from: MEMOIRS, Scientific Publications Department, The New York Botanical Garden, Bronx, NY 10458-5126 USA.

Memoirs of The New York Botanical Garden*

VOLUME		PRICE
46	**Mickel, L. T. and J. M. Beitel.** 1988. Pteridophyte flora of Oaxaca, Mexico. (568 pp.) hardbound	$90.00
47	**Park, C.-W.** 1988. Taxonomy of *Polygonum* section *Echinocaulon* (Polygonaceae). (82 pp.)	$19.00
48	**Samuels, G. J.** 1988. Fungicolous, lichenicolous, and myxomyceticolous species of *Hypocreopsis, Nectriopsis, Nectria, Peristomialis,* and *Trichonectria.* (78 pp.)	$18.50
49	**Samuels, G. J.** 1989. Mycological contributions celebrating the 70th birthday of Clark T. Rogerson. (375 pp.)	$74.00
50	**Renner, S. S.** 1989. Systematic studies in the Melastomataceae: *Bellucia, Loreya,* and *Macairea.* (112 pp.)	$24.00
51	**Maguire, B. and Collaborators.** 1989. The botany of the Guayana Highland—Part XIII. (127 pp.)	$28.75
52	**Thiers, B. M. and S. R. Gradstein.** 1989. Lejeuneaceae (Hepaticae) of Australia. I. Subfamily Ptychanthoideae. (82 pp.)	$17.50
53	**Proctor, G. R.** 1989. Ferns of Puerto Rico and the Virgin Islands. (389 pp.) hardbound	$79.50
54	**Delendick, T. J.** 1990. A survey of foliar flavonoids in the Aceraceae. (129 pp.)	$33.50
55	**Gottsberger, G. and G. T. Prance, Editors.** 1990. Reproductive biology and evolution of tropical woody Angiosperms. (195 pp.)	$40.50
56	**Dorr, L. J.** 1990. A revision of the North American genus *Callirhoe* (Malvaceae). (75 pp.)	$18.50
57	**Stevenson, D. W., Editor.** 1990. The biology, structure, and systematics of the Cycadales. (210 pp.)	$49.50
58	**Tiehm, A. and F. A. Stafleu.** 1990. Per Axel Rydberg: A biography, bibliography and list of his taxa. (75 pp.)	$19.00
59	**Samuels, G. J.** 1990. Contributions Toward a Mycobiota of Indonesia (180 pp.)	$36.25
60	**Mena V., P.** 1990. A Revision of the Genus *Arcytophyllum* (Rubiaceae: Hedyotideae). (26 pp.)	$ 7.25
61	**Grimes, J. W.** 1990. A Revision of the New World Species of Psoraleeae (Leguminosae: Papilionoideae). (113 pp.)	$23.25
62	**Barr, M. E.** 1990. Some Dictyosporous Genera and Species of Pleosporales in North America. (92 pp.)	$18.75
63	**Fulford, M. and A. J. Sharp.** 1990. The Leafy Hepaticae of Mexico: One Hundred and Twenty-seven Years after C. M. Gottsche (86 pp.)	$17.50
64	**Buck, W. R., B. M. Boom, and R. A. Howard, Editors.** 1990. The Bassett Maguire Festschrift: A Tribute to the Man and His Deeds. (302 pp.)	$59.75
65	**Barneby, R. C.** 1991. Sensitivae Censitae: A Description of the Genus *Mimosa* Linnaeus (Mimosaceae) in the New World. (835 pp.)	$

Postage and Handling:

U.S. ORDERS: $3.00 + 4% of subtotal.

NON-U.S. ORDERS: $4.00 + 6% of subtotal. Payment in U.S. currency drawn on a U.S. bank. Thank you.

Order from: MEMOIRS, Scientific Publications Department, The New York Botanical Garden, Bronx, NY 10458-5126 U.S.A.

* Volumes 1–45 have been published. Write to the address above for information as to their status and prices.

American Enterprise Institute for Public Policy Research
1150 Seventeenth Street, N.W. Washington, D.C. 20036 (202) 296-5616

AEI LEGISLATIVE ANALYSES

Legislative Analyses—95th Congress (to date)

Review—1977 Session of the Congress and Index of AEI Publications
Proposed Amendments to the National Labor Relations Act
Expanding the Right to Sue for Antitrust Violations
Full Employment and Balanced Growth Act: An Update
Hatch Act Revision
Bankruptcy Reform
Public Financing of Congressional Campaigns
Zero-Base Budgeting and Sunset Legislation
Federal No-Fault Insurance Legislation
Proposals for a National Court of Appeals
Military Base Closings: Benefits for Community Adjustment
Postal Service Legislative Proposals
Waterway User Charges
Electric Utility Rate Reform
The Renegotiation Reform Act of 1977
Criminalization of Payments to Influence Foreign Governments
Consumer Class Actions
Minimum Wage Legislation
Farm Policy Proposals
Direct Election of the President

Price: $2.00 per copy

AEI ASSOCIATES PROGRAM

The American Enterprise Institute invites your participation in the competition of ideas through its AEI Associates Program. This program has two objectives:

The first is to broaden the distribution of AEI studies, conferences, forums, and reviews, and thereby to extend public familiarity with the issues. AEI Associates receive regular information on AEI research and programs, and they can order publications and cassettes at a savings.

The second objective is to increase the research activity of the American Enterprise Institute and the dissemination of its published materials to policy makers, the academic community, journalists, and others who help shape public attitudes. Your contribution, which in most cases is partly tax deductible, will help ensure that decision makers have the benefit of scholarly research on the practical options to be considered before programs are formulated. The issues studied by AEI include:

- Defense Policy
- Economic Policy
- Energy Policy
- Foreign Policy
- Government Regulation
- Health Policy
- Legal Policy
- Political and Social Processes
- Social Security and Retirement Policy
- Tax Policy

For more information, write to: **American Enterprise Institute**
1150 Seventeenth Street, N.W.
Washington, D.C. 20036

NORTHERN ARIZONA UNIVERSITY

AMERICAN ENTERPRISE INSTITUTE
LEGISLATIVE ANALYSES

Air Transportation Regulatory Reform

39.3
45

THE AMERICAN ENTERPRISE INSTITUTE FOR PUBLIC POLICY RESEARCH, established in 1943, is a publicly supported, nonpartisan research and educational organization. Its purpose is to assist policy makers, scholars, businessmen, the press and the public by providing objective analysis of national and international issues. Views expressed in the institute's publications are those of the authors and do not necessarily reflect the views of the staff, advisory panels, officers or trustees of AEI.

Institute publications take four major forms:

1. **Studies**—in-depth studies and monographs about government programs and major national and international problems written by independent scholars.
2. **Legislative Analyses**—balanced analyses of current proposals before the Congress, prepared with the help of specialists in the fields of law, economics, and government.
3. **Forums and Conferences**—proceedings of discussions in which eminent authorities express contrasting views on public policy issues.
4. **Reviews**—timely presentations of informed opinion and information on the emerging issues of the day.

COUNCIL OF ACADEMIC ADVISERS

Paul W. McCracken, *Chairman, Edmund Ezra Day University Professor of Business Administration, University of Michigan*

Kenneth W. Dam, *Harold J. and Marion F. Green Professor of Law, University of Chicago Law School*

Milton Friedman, *Paul Snowden Russell Distinguished Service Professor of Economics, University of Chicago; Nobel Laureate in Economic Science*

Donald C. Hellmann, *Professor of Political Science and Comparative and Foreign Area Studies, University of Washington*

D. Gale Johnson, *Eliakim Hastings Moore Distinguished Service Professor of Economics and Provost, University of Chicago*

Robert A. Nisbet, *Albert Schweitzer Professor of Humanities, Columbia University*

G. Warren Nutter, *Paul Goodloe McIntire Professor of Economics, University of Virginia*

Marina v. N. Whitman, *Distinguished Public Service Professor of Economics, University of Pittsburgh*

James Q. Wilson, *Henry Lee Shattuck Professor of Government, Harvard University*

EXECUTIVE COMMITTEE

Herman J. Schmidt, *Chairman of the Board*
William J. Baroody, *President*
Charles T. Fisher III, *Treasurer*
Richard J. Farrell
Richard B. Madden
Richard D. Wood

William J. Baroody, Jr., *Executive Vice President*

Gary L. Jones, *Assistant to the President for Administration*
Edward Styles, *Director of Publications*

PROGRAM DIRECTORS

Russell Chapin, *Legislative Analyses*
Robert A. Goldwin, *Seminar Programs*
Robert B. Helms, *Health Policy Studies*
Thomas F. Johnson, *Economic Policy Studies*
Marvin H. Kosters, *Government Regulation Studies*
W. S. Moore, *Legal Policy Studies*
Rudolph G. Penner, *Tax Policy Studies*
Robert J. Pranger, *Foreign and Defense Policy Studies*